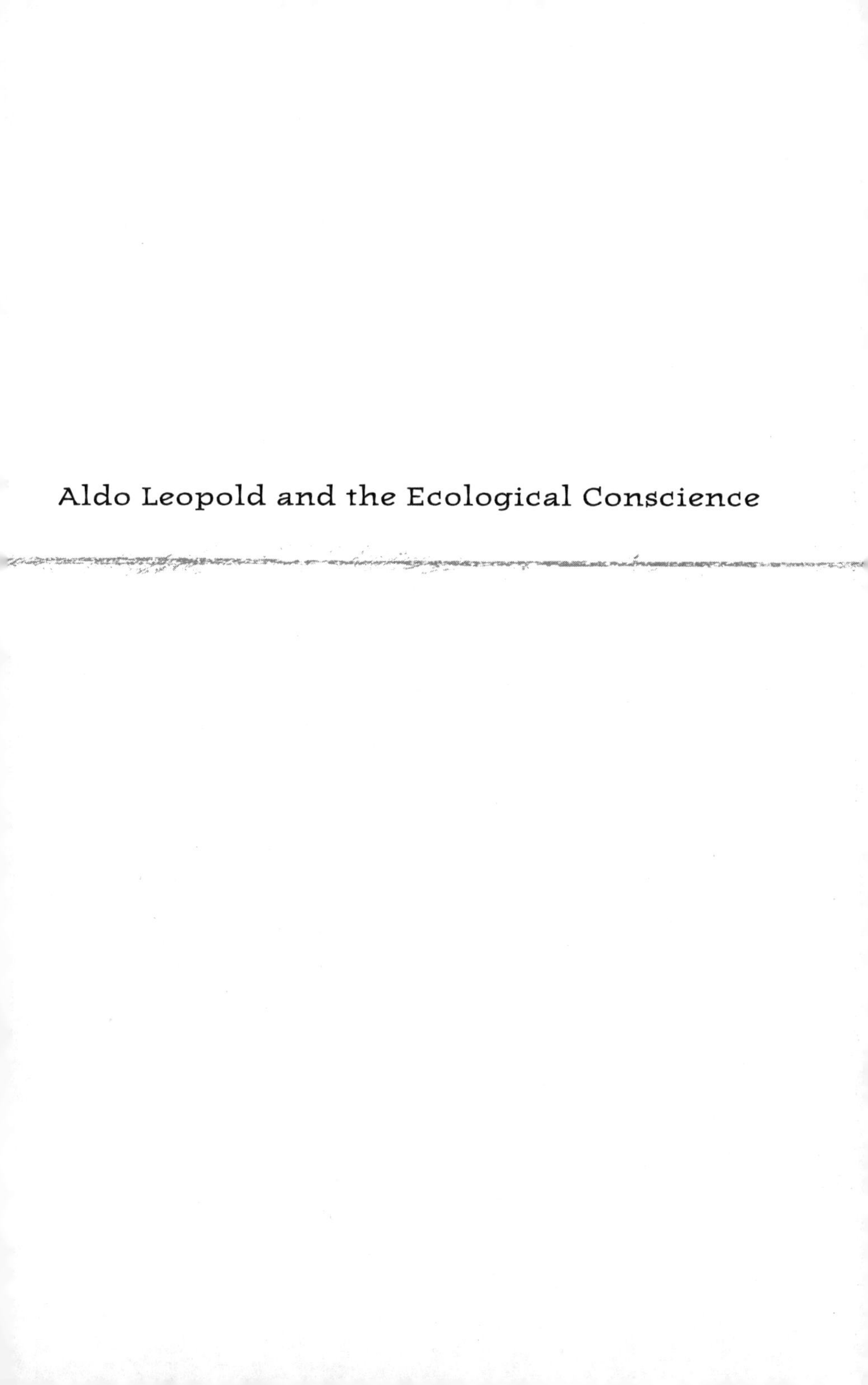

Aldo Leopold and the Ecological Conscience

Aldo Leopold and th

Ecological Conscience

EDITED BY RICHARD L. KNIGHT
AND SUZANNE RIEDEL

2002

Oxford New York
Auckland Bangkok Buenos Aires Cape Town Chennai
Dar es Salaam Delhi Hong Kong Istanbul Karachi Kolkata
Kuala Lumpur Madrid Melbourne Mexico City Mumbai Nairobi
São Paulo Shanghai Singapore Taipei Tokyo Toronto

and an associated company in Berlin

Published by Oxford University Press, Inc.
198 Madison Avenue, New York, New York 10016

www.oup.com

Library of Congress Cataloging-in-Publication Data
Aldo Leopold and the ecological conscience / edited by
Richard L. Knight and Suzanne Riedel.
p. cm.
Includes bibliographical references.
ISBN 0-19-514943-2; ISBN 0-19-514944-0 (pbk.)
1. Leopold, Aldo, 1886–1948—Influence. 2. Nature conservation.
I. Knight, Richard L. II. Riedel, Suzanne.
QH31.L618 A69 2002
333.7'2—dc21 2001036412

All photographs courtesy of the Aldo Leopold Foundation, Baraboo, WI,
with the exception of the photograph appearing on p. 60, which is by
Charles W. Schwarz. The title page photograph was taken by Robert McCabe.

9 8 7 6 5 4 3 2 1

Printed in the United States of America
on acid-free paper

To those who show solicitude for land health
and its ability to sustain human communities.

*I need a short name for what is lacking: I call
it the ecological conscience. Ecology is the science
of communities, and the ecological conscience
is therefore the ethics of community life.*

—Aldo Leopold,
"The Ecological Conscience," 1947

Acknowledgments

First and foremost, we wish to thank the essayists who took time from their already-too-busy schedules to contribute to our exploration of Aldo Leopold and his ecological legacy. By their commitment to conservation and their belief that Aldo Leopold represented something important in contemporary human–land relations, our essayists have allowed each of us to learn something about ourselves and our beliefs. The Aldo Leopold Foundation allowed us to use photos from their archives and enthusiastically supported the project from the onset. The cover photograph was taken by Robert McCabe, a recipient of the Wildlife Society's Aldo Leopold Award, student of Leopold, mentor, and friend. His family has graciously allowed us to use this picture as the book's cover. The Wildlife Society generously gave us permission to reprint those essays that originally appeared in a special issue of the *Wildlife Society Bulletin* honoring the half-century anniversary of *A Sand County Almanac*. (These are essays by Curt D. Meine, Richard L. Knight, Mary Anne Bishop, Reed Noss, Winifred B. Kessler and Annie L. Booth, Jamie Rappaport Clark, Edwin P. Pister, and Jack Ward Thomas.)

RLK wishes to thank the many people who have provided encouragement and support on his own journey through a delightful maze of people and land relationships. These include Stan and Junior Berg, Nina and Charlie Leopold Bradley, Wendell Gilgert, Joe Hickey, Robert Knight, Estella Leopold, Patty Limerick, Robert McCabe, Gary Meffe, Curt Meine, Phil Pister, Carroll Reick, Evan and Catherine Roberts, Stan Temple, George Wallace, a number of my graduate students at Colorado State University, my parents, and, most important, my wife Heather, the most remarkable land steward and caring person I have known. By naming me an Aldo Leopold Leadership Fellow, the Ecological Society of America, and particularly Jane Lubchenco, encouraged me in my studies of how ecology can become more relevant in contemporary society. The Aldo Leopold Foundation provided support early on in my studies of human and land health through a Leopold Fellowship. Charlie and Nina allowed me to live at the "shack" and provided me with meals, drinks, killer-diller kosher pickles, and insights about how to connect one's passion to a lifelong quest for healthy human and natural communities.

SR wishes to thank the friends and family members who provided encouragement and support, especially her husband Will, who contributed advice and artistic and technical help from the earliest stages. Edwin J. Jones, former editor of the *Wildlife Society Bulletin*, admirer and occasional impersonator of Aldo Leopold, and enthusiastic advocate of the land ethic in all of its many ramifications, deserves special thanks and appreciation.

Foreword

NINA LEOPOLD BRADLEY
AND WELLINGTON HUFFAKER

Through his own intellectual evolution, Aldo Leopold advanced the development of ecological science. He recognized ecology as the fusion point of science and the land community. Today, grounded in ecology, we have begun to understand that solving environmental issues necessitates connecting theory and facts across many disciplines and infusing our ecological knowledge with a sense of wonder and passion. The authors of these essays reveal the need to regain a sustainable relationship to place, community, and the natural world—to the land that supports all life.

The need for integration is a common theme among ecologists. In his important book *Consilience*, E. O. Wilson writes, "The greatest enterprise of the mind has always been, and always will be, the attempted linkage of the sciences and humanities. The ongoing fragmentation of knowledge and resulting chaos in philosophy are not reflections of the real world but artifacts of scholarship"(1998, 8). However, powerful trends in modern science, spurred by the growth of new technologies, lead away from integration of knowledge and toward ever greater specialization. The recent announcement of the successful mapping of the human genome is only one example of the increasing scientific focus on microscopic and submicroscopic levels of biological organization. As revealing of life as these are, knowledge without context is of little use in solving our major problems. In our rush to map genes, we seldom ask the question, Where will this map lead us? Will it lead us to more patents for the rising class of commercial biologists or to knowledge that benefits the biotic community as a whole? Ignoring relevant ethical questions may lead to outcomes we would not knowingly choose.

C. P. Snow, in his celebrated 1959 Rede Lecture "The Two Cultures and the Scientific Revolution," wrote that the polarization between the sciences and the humanities "is sheer loss to us all . . . it is at the same time a practical, intellectual and creative loss" (1959, 12). Throughout his life, Aldo Leopold expressed a similar concern. In 1935, during a visit to Germany, he described what he saw as unnaturally opposing forces in the intellectual community. Sitting in a Berlin hotel room one evening, he jotted down these thoughts on the back of a piece of hotel stationery:

> One of the anomalies of modern ecology is that it is the creation of two groups, each of which seems barely aware of the existence of the other. The one studies the human community almost as if it were a separate entity, and calls its findings sociology, economics and history. The other studies the plant and animal community and comfortably relegates the hodge-podge of politics to the 'liberal art.' The inevitable fusion of these two lines of thought will, perhaps, constitute the outstanding advance of the present century. (Leopold, unpublished notes)

Through his own participation in the land community, Leopold came to a deeper personal understanding and appreciation of the land. On his Wisconsin sand county farm, he struggled to rebuild a diverse, healthy, aesthetically satisfying biota on land degraded by generations of misuse. As he became more acutely aware of the complexity of factors involved in life and death, growth and decay, he developed a profound humility. Ethical and aesthetic values became entwined with his scientific understanding. Recording and integrating all the strands of his own firsthand experience, blending science with philosophy, history, literature and even poetry, he came to his final statement of the "land ethic": "That land is a community is the basic concept of ecology, but that land is to be loved and respected is an extension of ethics" (Leopold 1949, viii).

With his use of the words *loved* and *respected*, we see the integration of science with a broader humanism. In *A Sand County Almanac*, we hear both an intellectual and an emotional thread of consilience. We are also aware of a new and unusual sense of natural beauty beyond the conventionally "scenic," growing out of a sense of ecological and evolutionary awareness.

Aldo Leopold understood that conservation issues are not narrow and personal, but multidimensional—as broad as human population growth, climate change, and the global species extinction crisis, and as personal as the local economy, pollution in our backyards, and chemical residues in our food. He would be heartened by the emergence of fields such as environmental psychology, environmental ethics, ecological restoration, sustainable agriculture, conservation biology, and environmental history. Integrating science, history, and culture, the boundaries of these fields blur as we negotiate our way through the difficult environmental and conservation challenges that engulf our world. Leopold was familiar with the process; in 1934, he wrote these notes on German game management: "In the long run we shall learn that there is no such thing as forestry, no such thing as game management. The only reality is an intelligent respect for, and adjustment to, the inherent tendencies of land to produce life" (Leopold, unpublished notes).

In retrospect, it is now apparent why Leopold's writing rang true with his readers and with his own family, who participated in the ritual of caring, personal relationships that extended to include care of the land. By sharing his personal struggle to integrate knowledge and make informed and ethical decisions, he makes us aware of the need for an ethical connection to people and to places. As Kathleen Dean Moore has written, "love for places and love for humans nurture each other and sustain us all" (1999, 15–16).

Although Leopold could not have predicted the role he would play in the evolution of a conservation ethic, he understood the need for such an evolution:

> I have purposely presented the land ethic as a product of social evolution because nothing so important as an ethic is ever written . . . the evolution of a land ethic is an intellectual as well as emotional process. Conservation is paved with good intentions which prove to be futile, or even dangerous because they are devoid of critical understanding either of the land, or of economic land use. I think it is a truism that as the ethical frontier advances from the individual to the community, its intellectual content increases. (Leopold 1949, 225)

The authors of the essays in this volume demonstrate that this evolution is ongoing. As educators, scholars, and practitioners, they represent the richness of the Leopold legacy. They encourage us to resist the path of specialization and isolation in contemporary science and, instead, to find our own path, developing and listening to our own ecological conscience, and remembering the context for all decisions—the land itself and all of us on the land.

References

Leopold, A. 1949. *A sand county almanac and sketches here and there.* New York: Oxford University Press.

Moore, K. D. 1999. An ethic of care. *Inner Voice* March/April:15–16.

Snow, C. P. 1959. *The two cultures and the scientific revolution.* New York: Cambridge University Press.

Wilson, E. O. 1998. *Consilience: The unity of knowledge.* New York: Alfred A. Knopf.

Contents

Contributors

Mary Anne Bishop is a wildlife research biologist with the Prince William Sound Science Center of Cordova, Alaska. She received her Ph.D. in wildlife ecology from the University of Florida. Her current research interests include shorebird migration ecology; predator–prey relations of shorebirds and gulls; and the conservation and management of cranes, waterfowl, and freshwater wetlands. Her research in Alaska is conducted on the Copper River Delta and in Prince William Sound. Since 1987, Dr. Bishop has worked cooperatively with the International Crane Foundation and Chinese scientists on the Tibetan Plateau studying the ecology and conservation of bar-headed geese and the endangered black-necked crane.

Annie L. Booth is an associate professor in the Environmental Studies Program at the University of Northern British Columbia. She received her B.A. in history from the University of Victoria, M.S. in environmental studies from York University, and Ph.D. in land resources from the University of Wisconsin–Madison. Her previous positions include: researcher for the Wisconsin Department of Natural Resources, research associate with the University of Wisconsin Cooperative Extension Service, and executive director of the Quetico Foundation in Toronto, Ontario. Her current research focuses on environmental ethics, resource-dependent communities, and relationships of Canada's First Nations to natural resources.

Nina Leopold Bradley, eldest daughter of Aldo Leopold, has undertaken ecological research throughout her life, has established two family planning clinics (Columbia, Missouri, and Bozeman, Montana), and currently lectures widely on Leopold and the land ethic. She and her husband, Charles, have directed research and ecological restoration at the Leopold Memorial Reserve since 1978. She received an honorary doctorate degree in environmental sciences from the University of Wisconsin in 1988 and received the Wilderness Society's Bob Marshall Award in 1995.

J. Baird Callicott is professor of philosophy and religion studies at the University of North Texas. He is author or editor of many books and articles about Aldo Leopold, including *The River of the Mother of God and Other Essays by Aldo Leopold*; *Companion to A Sand County Almanac: Interpretive and Critical Essays*; *In Defense of the Land Ethic*; *Essays in Environmental Philosophy*; *Beyond the Land Ethic: More Essays in Environmental Philosophy*; and *Aldo Leopold: For the Health of the Land.*

Jamie Rappaport Clark is the former director of the U.S. Department of Interior's Fish and Wildlife Service. She currently is Senior Vice President for Conservation Programs in the National Wildlife Federation. She received her B.S. in wildlife biology from Towson State University and her M.S. in wildlife ecology from the University of Maryland–College Park, where her graduate work focused on white-tailed deer. She lives in Leesburg, Virginia, with her husband, Jim, and their son Carson.

James A. Estes is a wildlife research biologist for the Western Ecological Research Center, Biological Resources Division, U.S. Geological Survey, and an adjunct professor at the University of California–Santa Cruz. He received his Ph.D. from the University of Arizona. He is a Pew Fellow in marine conservation and fellow of the California Academy of Sciences. He has served on the editorial boards of *Ecology*, *Ecological Monographs*, *Animal Conservation*, and *Marine Ecology Progress Series*. Dr. Estes is interested in predation as an ecosystem-level process and has studied sea otters and their kelp forest habitats since 1970.

Wellington "Buddy" Huffaker was hired as the Aldo Leopold Foundation's first ecologist in 1996 and appointed Executive Director in 1999. Under his leadership, ALF has expanded both its national programs increasing the awareness and appreciation for the legacy of Aldo Leopold as well as its regional programs designed to educate, inspire, and empower private landowners in south-central Wisconsin to undertake ecological management of their properties.

Stephen R. Kellert is a professor at the Yale University School of Forestry and Environmental Studies, where he studies the connection between human and natural systems. His awards include the National Conservation Achievement Award (National Wildlife Federation) and the Distinguished Individual Achievement Award (Society for Conservation Biology). He has authored over one hundred publications, including the following books: *Kinship to Mastery: Biophilia in Human Evolution and Development*; *The Value of Life: Biological Diversity and Human Society*; *The Biophilia Hypothesis* (with E. O. Wilson); and *Ecology, Economics, Ethics: The Broken Circle* (with F. H. Bormann).

Winifred B. Kessler is the director of Wildlife, Fisheries, Ecology and Watershed for the USDA Forest Service in the Alaska region. Formerly, she was professor and chair of the Forestry Program at the University of Northern British Columbia. She obtained her Ph.D. from Texas A&M University. She joined The Wildlife Society in 1975 and became a Certified Wildlife Biologist in 1978. Her career spans a variety of teaching, research, and management positions. During the early 1990s, she (with Hal Salwasser) directed the USDA Forest Service "New Perspectives" program, which generated the concept of ecosystem management, the organizing approach for so many contemporary natural resource organizations. She is an active volunteer in conservation work, including service as a professional member of the Boone and Crockett Club.

Richard L. Knight is a professor of wildlife conservation at Colorado State University. He received his Ph.D. from the University of Wisconsin, where he was a Leopold Fellow. He is co-editor of *The Essential Aldo Leopold: Selected Quotations and Commentaries*, published by the University of Wisconsin Press. Dr. Knight was recently selected as an Aldo Leopold Leadership Fellow by the Ecological Society of America. He has co-edited *Wildlife and Recreationists*; *A New Century for Natural Resources Conservation*; *Stewardship Across Boundaries*; *Forest Fragmentation in the Southern Rocky Mountains*; and *Ranching West of the 100th Meridian*. With his wife, Heather, he practices community-based conservation in the Livermore Valley, Colorado.

L. David Mech is a wildlife research biologist for Northern Prairie Wildlife Research Center, Biological Resources Division, U.S. Geological Survey, and an adjunct professor at the University of Minnesota. He has a B.S. degree from Cornell University and a Ph.D. from Purdue University. Since 1958, he has studied wolves and their prey and various other carnivores. He is the author of *The Wolf: The Ecology and Behavior of an Endangered Species* and co-author of *The Wolves of Denali*. He is a recipient of The Wildlife Society's Aldo Leopold Award.

Curt D. Meine is a research associate with the International Crane Foundation. He received a B.A. in English and history from DePaul University and an M.S. and a Ph.D. in land resources from the University of Wisconsin–Madison. He has served as a consultant to many local, national, and international conservation agencies and organizations, and he has taught at the University of Wisconsin–Madison. He is author of the biography *Aldo Leopold: His Life and Work*; co-compiler of *The Cranes: Status Survey and Conservation Action Plan*; editor of *Wallace Stegner and the Continental Vision: Essays on Literature, History, and Landscape*; and co-editor of *The Essential Aldo Leopold: Selected Quotations and Commentaries*. Dr. Meine has contributed to a wide variety of periodicals, journals, and books, and he currently serves on the editorial boards of *Conservation Biology* and *Environmental Ethics* and on the Board of Governors of the Society for Conservation Biology.

Reed Noss is chief scientist for Conservation Science, Inc., an international consultant and lecturer, former president of the Society for Conservation Biology, former editor of *Conservation Biology*, science editor of *Wild Earth*, and a martial arts instructor. He is the author of nearly two hundred scientific and semitechnical articles and five books, including *Saving Nature's Legacy* and *The Science of Conservation Planning*. He lives with his family in the foothills of the Oregon Coast Range, outside of Corvallis.

Edwin P. Pister retired in 1990, following thirty-eight years as a fishery biologist with the California Department of Fish and Game. He studied wildlife conservation and zoology under A. Starker Leopold at the University of California–Berkeley and has spent virtually his entire career supervising aquatic management and research within an area encompassing approximately one thousand waters of the eastern Sierra and desert regions of California, ranging from the crest of the Sierra Nevada to the floor of Death

Valley. He founded and serves as executive secretary of the Desert Fishes Council and is involved in arid land ecosystem preservation throughout the American Southwest and adjoining areas of Mexico. He holds special interest in the fields of conservation biology and environmental ethics. He serves on the Board of Governors of the American Society of Ichthyologists and Herpetologists and as an emeritus member of the governing board of the Society for Conservation Biology. He also serves on the President's Advisory Committee of the University of California's systemwide White Mountain Research Station. He has lectured at more than seventy universities in North America and the United Kingdom and has authored seventy-six published papers and book chapters.

Suzanne Riedel received her Master of Fisheries and Wildlife Science degree from North Carolina State University. She holds an M.A. in English from the University of Arizona and has worked in publishing and editing for a number of years, most recently as production manager and special feature editor for the *Wildlife Society Bulletin.* She is currently employed as a research assistant to the curator of research at the North Carolina Zoological Park.

John Seidensticker is senior curator at the National Zoological Park and author of over 140 articles and books, including *Saving the Tiger* and *Riding the Tiger: Tiger Conservation in Human-Dominated Landscapes.* He pioneered the use of radiotelemetry to study the mountain lion in North America and was founding director of the Smithsonian-Nepal Tiger Ecology Project. He has traveled widely in Asia and served as an ecologist and park planner for the Indonesia World Wildlife Fund Program. He has conducted fieldwork in the Sundarbans of Bangladesh and India, in Thailand, and in Sri Lanka. As a conservation biologist at the Smithsonian National Zoological Park, Dr. Seidensticker's research efforts have focused on understanding and encouraging landscape patterns where large mammals can persist, training future conservation leaders, and diffusing environmental understanding through his writing, public appearances, and museum and zoo exhibits.

Jack Ward Thomas is the Boone and Crockett Professor in the School of Forestry at the University of Montana. He served ten years as a wildlife biologist for the Texas Parks and Wildlife Department, twenty-seven years as a research biologist, and four years as chief of the USDA Forest Service. He received a B.S. in wildlife management from Texas A&M University, an M.S.

in wildlife ecology from West Virginia University, and a Ph.D. in forestry from the University of Massachusetts. He has authored over 350 publications, ranging from ungulate ecology to ecosystem management. He has served as president of The Wildlife Society and is a recipient of the Aldo Leopold Award.

Aldo Leopold and the Ecological Conscience

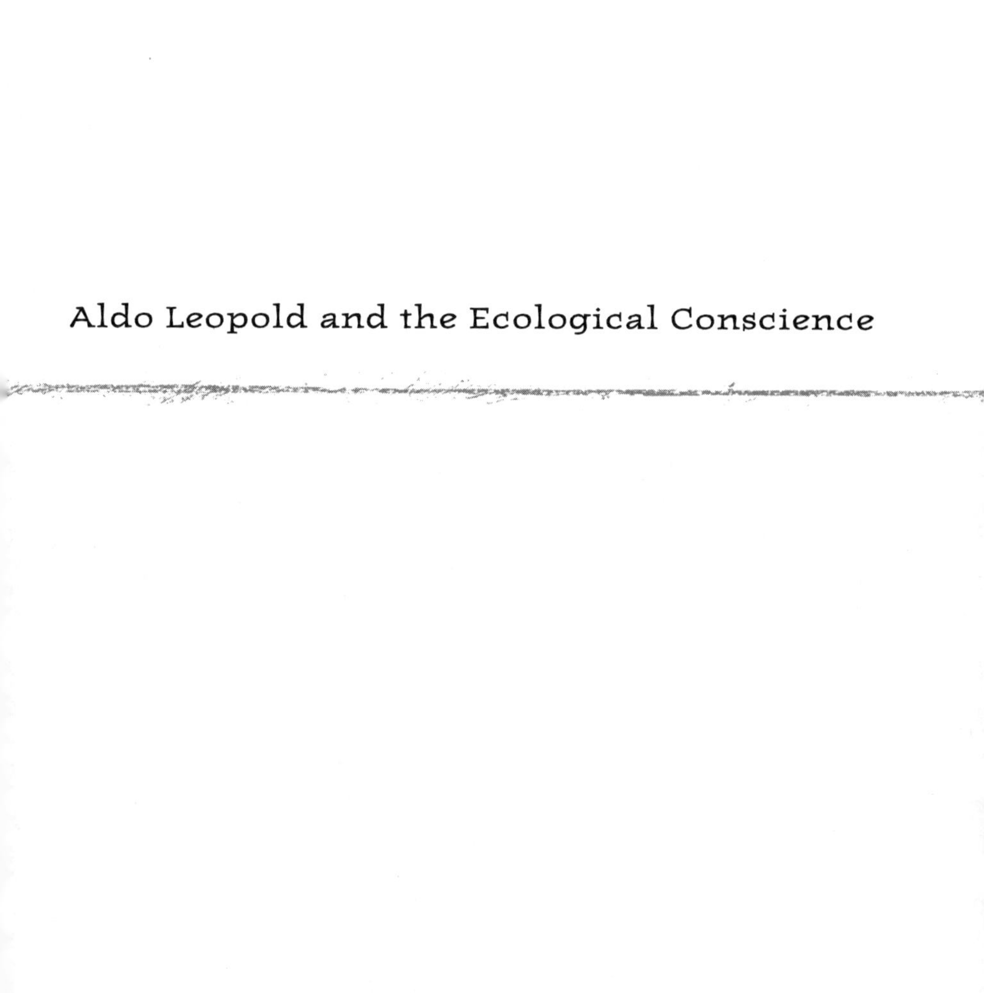

Introduction

RICHARD L. KNIGHT
AND SUZANNE RIEDEL

Aldo Leopold is most widely known and appreciated for his description of the land ethic, as put forth in his collection of essays, *A Sand County Almanac*. Leopold's land ethic has been interpreted by readers and scholars to mean different things, but at the very least, most agree it was Leopold's attempt to describe a responsible, meaningful relationship between individuals and the land (Callicott and Freyfogle 1999). In the history of natural resources conservation, seldom has there been a more urgent need than now for scientists and land stewards with the far-reaching vision and shrewd ecological insights that Aldo Leopold possessed. Thanks to *A Sand County Almanac*, today society has more, rather than fewer, individuals of this kind to draw upon. Fifty years after its publication, *A Sand County Almanac* is still attracting new recruits into the field of conservation—young scientists and land managers determined to put right the damage done from the past century of exploitation.

This collection of essays by ecologists, wildlife biologists, and conservationists documents the legacy of Aldo Leopold and *A Sand County Almanac* to the environmental movement, ecological sciences, and natural resource management. For some of these authors, Leopold's land ethic and the interdisciplinary approach he exemplified have been crucial for coming to grips with many of the large, numbingly complex problems confronting science today. Jamie Rappaport Clark describes how the U.S. Fish and Wildlife Service integrated Leopold's land ethic into its daily operations as part of its strategy to "think globally and act locally." Jack Ward Thomas recounts how *A Sand County Almanac* influenced him to write a document of land-use ethics to guide Forest Service personnel in the management of 194 million acres of public land.

But the impact of Leopold's philosophy and of his code of ethics among scientists may be far greater than generally supposed. For many, Leopold's influence has been personal, one on one (author to individual reader). The truth of Leopold's message in *A Sand County Almanac* may be generally acknowledged, but because of the unconventional style of the book (particularly in its use of language that is emotive, symbolic, and intuitive), Leopold may not have received the credit he deserves as an instigator of major changes in the management of natural resources.

Ironically, the very reasons for the success of *A Sand County Almanac* may undermine its credibility among some scientists. As Curt Meine argues in our book's opening essay, *A Sand County Almanac* was intended to be an example of the kind of ecological thinking Leopold advocated—integrative, grounded in a sense of history and experience, emphasizing process over products, and bringing intuition and other modes of perception into play. In his own day, Leopold observed a yawning gap emerging between the "complexity of the land organism" being revealed by the new ecology and the inadequacies of conventional science to explain it. Scientists, then, needed to expand their skills beyond conventional methodologies, to fine-tune their observations through direct experience, intuitions, and even empathy. In writing *A Sand County Almanac*, Leopold used his considerable powers of observation and, ignoring "the senseless barrier between science and art," created a timeless classic.

Although *A Sand County Almanac* may not have received the full professional acknowledgment it deserves, it has succeeded in what Leopold intended first and foremost—to influence the way ecologists think. For example, in a survey of natural resources managers asked to list the three most important sources of information to their professional careers, Leopold was listed by 90% of the participants. Furthermore, to those who oppose the shifts in thinking and management practices that have come about with ecosystem management, Leopold's early and continuing influence has been plainly evident. Several years ago, in a debate on Leopold's legacy in the *Journal of Forestry*, Boris Zeide stated bluntly, "Leopold is the spiritual father of ecosystem management, and his legacy remains at the center of contemporary issues in forestry." And he added that "Leopold's influence is based largely on a brief essay (20-odd pages) that outlines what he calls the land ethic. This essay comes near the end of *A Sand County Almanac*, a collection of charming nature sketches" (Zeide 1996, 13–14). Among the many rejoinders to Zeide's critique of the land ethic, the most cogent was by Baird Callicott, who called attention to the fact that history has voted in Leopold's favor; the tide has turned and "the balance of power is fixing to shift" (Callicott 1996, 26). Leopold's influence seems to have prevailed, in some important quarters, at least for the moment.

These essays may explain why. In acknowledging a debt of gratitude from the scientific community, our authors call attention to some of the many values of *A Sand County Almanac* for ecologists and other professionals engaged in the work of conservation. Among these values are the following:

(1) A standard of professional conduct that is challenging and capable of effecting change. *A Sand County Almanac* offers a clear alternative to the status quo, urging ecologists and other scientists to extend their efforts beyond their fields of expertise, to take risks in areas where knowledge is limited or uncertain. It also proposes a standard of commitment (personal, ethical, and political). Leopold saw this as a practical concern because he thought it unlikely that, without a level of personal commitment, scientists would have any real impact on society at large. His ironic definition of conservation was "a series of ecological predictions made by beginners because ecologists have failed to offer any" (Callicott and Freyfogle 1999, 220). Leopold believed that scientists should have the courage, persistence, and commitment necessary to follow the truth of their convictions, wherever that might lead.

(2) A source of fresh insights and ideas. As many of these essays attest, *A Sand County Almanac* is a book of ideas that can be read again and again for new insights. Those who think they know the book well might consider a return visit. Career professionals returning to *A Sand County Almanac* often discover a different book from the one they read twenty years earlier—one that is pragmatic, tough-minded, and relevant to the issues of science today.

As Winifred Kessler and Annie Booth point out in their essay "Professor Leopold, What is Education For?" *A Sand County Almanac* seems to have something for everyone, and people tend to pick and choose to suit their needs. Another way of looking at this is to think of the text as multilayered; it tends to reward experience. In "What Would Aldo Have Done?" Jack Ward Thomas describes the process of rereading *A Sand County Almanac*: "In the course of that latest visit, I discern things in the written landscape that I had not sensed before. This, I believe, is the result of things that I have learned . . . Colleagues have expressed similar feelings to me."

This raises the question, How many of these ideas and insights did Leopold himself intend? None? All? Is *A Sand County Almanac* merely a handful of charming nature sketches, as some critics contend, or a dense parable, full of hidden meaning, as literary analysis suggests? ("The smoothness and density of Leopold's prose belies its density. Like hand-rubbed wood, its surface conceals its craft"; Tallmadge 1987, 115.) In either case, *A Sand County Almanac* remains a fertile ground for thought. Leopold had talent as a writer, and apparently also as a teacher, for actively engaging and challenging an audience. He told his wildlife biology students, "I am trying to teach you that this alphabet of 'natural objects' (soils and rivers, birds and beasts) spells out a story, which he who runs may read—if he knows how" (Flader and Callicott 1991, 337).

(3) A holistic, interdisciplinary approach. This approach emphasizes the importance of every cog and wheel in maintaining ecosystems. Leopold's ability to synthesize information from many sources and his intuitive understanding of the complexity of ecosystems led him to take a conservative approach to land management. His extensive reading of history had instilled in him a sense of humility and caution. He advocated that management decisions be undertaken in the widest possible context, with full consideration of the evolutionary history and interrelationships of an organism or place.

(4) An answer to cynicism. As Leopold said, "One of the penalties of an ecological education is that one lives alone in a world of wounds" (Leopold 1953, 165). *A Sand County Almanac* offers an embracing vision that places high demands on the enlightened individual as the important unit of conservation effort. Thinking like a mountain is not a characteristic of institutions. Furthermore, it is not a result of knowledge, aesthetic realization, or instinct alone. Leopold described it in *A Sand County Almanac* as a process of social evolution that begins as a synthesis of many modes of perception in individuals and is perpetuated in the community at large because of its survival value. Those who embrace the land ethic may be, in Reed Noss's words,

"a tiny minority," but they have the satisfaction of being in the vanguard of social evolution, partaking in a uniquely human capacity. Leopold described the capacity of one species to care about the fate of another as "a new thing under the sun" (Leopold 1987, 110).

As an ecologist, Leopold makes an intellectual appeal for understanding that human beings are not, after all, monarchs, but plain citizens of the biotic community. At the same time, Leopold the conservationist and reformer makes a more energetic and urgent appeal: human beings occupy a position of unique privilege and authority among all other species; they are the arbiters of life and death, capable of changing the face of the earth. They should, therefore, acknowledge their role in evolutionary history and act accordingly—with a sense of high purpose and restraint. In "The Land Ethic," Leopold makes an appeal for future generations, calling on Arthurian legend and Robinson's injunction to Tristram to "Mark what you leave" (Leopold 1987, 223). Leopold himself saw no point in understating a challenge. "In our attempt to make conservation easy, we have made it trivial" (Leopold 1987, 210).

(5) An effective antidote to utilitarian ideologies. These typically are the norm in natural resources management. Many of the essayists here have found that *A Sand County Almanac* provides an important corrective influence in opposition to dehumanizing mechanisms in business, government, and educational institutions. From the standpoint of science in the public arena, former Secretary of Interior Bruce Babbitt has surmised, "In our time the land ethic is still the crucial short-term moral counterforce in opposition to expediency in determining the use of our land, air, and water resources" (Babbitt 1987, 138). In "The A-B Dichotomy and the Future," Phil Pister describes how, after a decade of field experience, he found in *A Sand County Almanac* "a rational basis for approaching and solving the problems that had perplexed and overwhelmed me. I felt I had within my grasp the basic components for making management programs address the entire biota, not simply the popular demands for fishing."

(6) The example of a life. Finally, there is the character of Aldo Leopold himself, an eloquent example of the land ethic in practice. Leopold presents the "delights and dilemmas" of the experiences that shaped and changed his thinking as an ecologist. For many readers, particularly those engaged in fieldwork, the acquaintance with Leopold is one of the greatest benefits of the book.

Albert Hochbaum, an avian ecologist and onetime student of Leopold's, called the essays, which he had been sent to review, "a self-portrait." He explained, "This series of sketches brings the man himself into focus. . . . It tells not what is law and order in his field, as most of his other writings have, but shows the process of his thinking" (Meine 1988, 457). In the original foreword (1947) to the collection of essays, then titled *Great Possessions,* Leopold acknowledged the autobiographical nature of the book: "I do not imply that this philosophy of land was always clear to me. It is rather the end-result of a life-journey, in the course of which I have felt sorrow, anger, puzzlement, or confusion over the inability of conservation to halt the juggernaut of land-abuse. These essays describe particular episodes, en route" (Leopold 1947, 2).

Reed Noss describes Leopold as a "parental figure to wildlife biologists and managers, as well as to foresters," adding that he is "as admired by these professionals today as he was in his lifetime." It is surprising how often in our book's essays Leopold is described as a mentor, friend, and companion. It is a common conceit among admirers of *A Sand County Almanac* to imagine themselves on a walk with Aldo Leopold. When Leopold wrote, "We can be ethical only in relation to something we can see, feel, understand, love, or otherwise have faith in," he was thinking of (hoping for) a cultural transformation. What he may not have anticipated was that, for many travelers still on the way to ecological enlightenment, the ethical connection to nature they could "see, feel, understand, love, or otherwise have faith in" would be found in the narrator himself.

Leopold spent the last fifteen years of his career doing what he thought of as laying the groundwork of conservation—teaching. His preferred method was by example and simple demonstration, leaving students to gather their own meaning. As he admitted to a colleague, "I am not sure at all that I am any good at mass production in students" (Meine 1988, 514). But Leopold was a dedicated teacher who found the one-on-one milieu he preferred as a writer.

The lesson of *A Sand County Almanac* is as simple, straightforward, and accessible as Leopold could make it. An ethic grounded in a privileged lifestyle would not have served his purpose. It was important that the site he chose for a demonstration of the land ethic was essentially worthless as "property." On his sand county farm, Aldo Leopold, family, and friends deliberately set an example that others could follow. The requirements to be a conservationist were simple: work hard and develop a set of principles to live by.

In parts I and II of *A Sand County Almanac*, the rewards of such a life unfold in the author's narrative of daily routines at the shack and in the adventures and revelations of his experiences as a hunter. These examples compose the text out of which the land ethic in part III emerges. Like Thoreau's cabin on Walden Pond, Leopold's shack on the farm is the focal point of an important experiment—to test the value of a life lived in a close and harmonious relationship to the land. And, like Thoreau's *Walden*, Leopold's *A Sand County Almanac* is the published report. Both books broke new intellectual ground and drew on a tradition of democratic values—hard work, ingenuity, nonconformity, a distaste for excess of all kinds, and a spirit of adventure.

Thoreau, however, chose to distance himself from Concord society. "I love to be alone," he said; "I never found the companion that was so companionable as solitude" (*Walden,* chapter V). In contrast, Leopold had a wife, family, and friends, who visited the shack to take part in the experiment. Solitude was not what he sought—it did not fit his democratic model. Thoreau advised, "Live free and uncommitted. It makes but little difference whether you are committed to a farm or the county jail" (*Walden,* chapter II). But as Leopold knew, most men and women are not free in Thoreau's sense.

Today, conservationists, professionals, and laymen alike have come to realize, as Leopold did, that saving the land for future generations is a social experiment. The model of a conservationist's life presented in *A Sand County Almanac* is not a voice crying out in the wilderness, but a man with his family and friends restoring a farm. Leopold describes his purpose explicitly in the foreword: "On this sand farm in Wisconsin, first worn out and then abandoned by our bigger-and-better society, we try to rebuild, with shovel and axe, what we are losing elsewhere. It is here we seek—and still find—our meat from God" (Leopold 1987, viii).

This is an example that speaks to our time, especially to Americans. The physical frontier has receded, but the qualities of character it produced, and that have been such an important part of our culture, are still in demand.

Acknowledging Leopold's Legacy

The progress of conservation is slow, as Leopold was the first to acknowledge. "The job we thought would take five years will barely be started in fifty" (Leopold 1953, 145). However, Leopold was an optimist when it came to people. He was encouraged by the new recruits he saw coming into the profession: "Ill-trained, many of them; intellectually tethered by bureaucratic superiors, most of them; but in dead earnest, nearly all of them . . . hungry to learn new cogs and wheels, eager to build a better taste in natural objects" (Leopold 1953, 154). Leopold's friend, Albert Hochbaum, described Leopold's unbounded optimism for the future as one of his strongest characteristics. It was an optimism based, not on his perception of the state of things as they were, but on his faith in people and their capacity for change. As he said in "The Land Ethic," "Only the most superficial student of history supposes that Moses 'wrote' the Decalogue; it evolved in the minds of a thinking community, and Moses wrote a tentative summary of it for a 'seminar.' I say tentative because evolution never stops" (Leopold 1987, 225).

The essays in this volume provide testimony that Leopold's optimism was justified. The near universal opinion about Leopold the man—held by those who knew him—was that he brought out the best in people. These essays assert Leopold's gift as the writer of *A Sand County Almanac* to do the same. Reed Noss states simply what seems to be the common sentiment among these writers: "It took a rereading or two of Leopold to teach me that natural science can help us learn how to live."

This book may contribute to an understanding of the role *A Sand County Almanac* has played in raising the stakes for and potential rewards of science practiced in the cause of conservation. We hope it will also contribute to a tradition among some scientists, like Jack Ward Thomas, of passing on the message of the land ethic: "And, as Charlie Wallmo did for me in class some forty-three years ago, I place a copy of *A Sand County Almanac* in the students' hands and ask for their reviews. These reviews differ from those made in my class so long ago. . . . It is, I think, a sign of progress and maturity in our profession—students beginning to 'think like a mountain.'"

References

Babbitt, B. 1987. The land ethic: a guide for the world. In *Aldo Leopold: The man and his legacy*, ed. T. Tanner, 137–144. Ankeny, Iowa: Soil Conservation Service of America.

Callicott, J. B. 1996. A critical examination of "Another Look at Leopold's Land Ethic." *Journal of Forestry* 96:20–26.

Callicott, J. B., and E. T. Freyfogle, eds. 1999. *Aldo Leopold. For the health of the land: Previously unpublished essays and other writings*. Washington, D.C.: Island Press.

Flader, S. L., and J. B. Callicott, eds. 1991. The River of the Mother of God *and other essays by Aldo Leopold*. Madison: University of Wisconsin Press.

Leopold, A. 1947. Foreword to *Great Possessions*, 31 July 1947, Leopold Papers 6B16, University of Wisconsin–Madison Archives.

Leopold, A. 1953. *Round River: From the journals of Aldo Leopold*. L. B. Leopold, ed. New York: Oxford University Press.

Leopold, A. 1987. Special commemorative edition of *A Sand County Almanac and Sketches Here and There*. New York: Oxford University Press.

Meine, C. D. 1988. *Aldo Leopold: His life and work*. Madison: University of Wisconsin Press.

Tallmadge, J. 1987. Anatomy of a classic. In *Companion to* A Sand County Almanac: *Interpretive and critical essays*, ed. J. B. Callicott, 110–127. Madison: University of Wisconsin Press.

Zeide. B. 1996. Another look at Leopold's land ethic. *Journal of Forestry* 96:13–19.

The Evolution of a Classic

Chapter 1

Moving Mountains

Aldo Leopold and *A Sand County Almanac*

His love was for present things, and these things were present somewhere; to find them required only the free sky, and the will to ply his wings. *Aldo Leopold*

CURT D. MEINE

Literary classics are the mountains of our minds. They shape us, subtly and continually. They cast long shadows. They provide access to higher realms. They make their own intellectual weather. We are prone to take them for granted, yet they so define our view of the world, and of ourselves, that we can hardly imagine the world without them.

The history of American conservation contains its own range of classics: Thoreau's *Walden,* Marsh's *Man and Nature,* Muir's *My First Summer in the Sierra,* Stegner's *Beyond the Hundredth Meridian,* Carson's *Silent Spring,* Abbey's *Desert Solitaire*, and Berry's *The Unsettling of America,* to name just a few of the high peaks. But conservation's literary landscape is rich in its variety and abundance, and it holds many less prominent but no less durable expressions. We return to their pages again and again, and we always find in them something timeless and something new.

Because mountains seem permanent, we may easily disregard the intense tectonic shifts and internal pressures that gave rise to them. Behind every story is another story. *A Sand County Almanac* is no exception. The face of this particular mountain is still fresh. The very pebbles still seem to be settling. *A Sand County Almanac* emerged from economic depression, the Dust Bowl, and World War II, in a time of rapid technological change, scientific revolutions, and widespread environmental deterioration. But it was also a time when its author, Aldo Leopold, strove to provide solid foundations for conservation (Meine 1994). His conviction was that conservation had to rest on a base that included not only the integrated natural sciences, but also philosophy, ethics, history, and literature. *A Sand County Almanac,* as it turned out, was the final proof of his conviction.

The State of the Profession, circa 1940

Future generations of conservationists would have understood if, by 1940, Aldo Leopold had begun to rest on his laurels, which were many. Building on his early years as an innovative forester in the U.S. Forest Service, Leopold had emerged in the 1930s as one of the preeminent leaders in wildlife ecology and management (Flader 1974; Meine 1988). Those years saw extraordinary changes in the field, and Leopold was involved in most of them (Dunlap 1988, Meine 1988). In the late 1920s, under contract with the Sporting Arms and Ammunitions Manufacturers' Institute, he had begun the field studies whose results appeared in his landmark *Report on a Game Survey of the North Central States* (Leopold 1931). Increasingly attentive to the fundamental relevance of ecology, Leopold argued in print and at the podium that game management had to refocus its efforts away from the tighter hunting regulations, game farming operations, and predator-control programs that then dominated game conservation, and toward the protection, enhancement, and restoration of habitat. This essential shift in approach gained the imprimatur of the American Game Association in its "American Game Policy," a document (destined for a long life) prepared by a committee that Leopold chaired (Leopold 1930). With publication of Leopold's *Game Management* in 1933, the field gained its first textbook; with Leopold's appointment that same year to the Chair of Game Management at the University of Wisconsin, it gained its first full-time professor. Leopold's students from the prewar period provided many of the profession's early leaders, including Frances Hamerstrom, Frederick Hamerstrom, Arthur Hawkins, Joe Hickey, H. Albert Hochbaum, Robert McCabe, and Lyle Sowls (McCabe 1988). Through the 1930s, Leopold helped to found or to redirect many of the nation's leading conservation agencies and nongovernmental organizations, including the U.S. Soil Conservation Service, the U.S. Biological Survey, the

U.S. Forest Service, the Society of American Foresters, the Wilderness Society, and the National Wildlife Federation.

Within wildlife management proper, Leopold played a central role in the formative events of the mid to late 1930s: creation of the American Wildlife Institute (later the Wildlife Management Institute) and development of the Cooperative Wildlife Research Unit system in 1935, establishment of the North American Wildlife Conference in 1936, founding of The Wildlife Society and its *Journal of Wildlife Management* in 1937, and passage and implementation of the Pittman-Robertson Federal Aid in Wildlife Restoration Act of 1937 (Trefethen 1975; Meine 1988). Important as these events were, they were only outward indicators of still deeper changes in wildlife conservation. The field was expanding to encompass not only game species, but also nongame animals and (at least for some managers) wild plants. At the same time, concern for threatened species and biotic communities was increasing (Leopold 1936). Wildlife management was securing an academic foothold in both basic and applied sciences, particularly in field ecology (Leopold 1939). Its practitioners were beginning to explore methods of restoring native species and communities. Its theoreticians were beginning to develop more sophisticated connections with colleagues in education, the agricultural sciences, economics, and public policy.

According to Flader and Callicott (1991, 19), "By 1940, Leopold could survey from its pinnacle the profession he had done more than anyone else to create." He served as president of the young Wildlife Society that year and used the opportunity of his presidential address in March 1940 to step back and put the recent advances into perspective. Leopold's message in "The State of the Profession" was at once practical and visionary, sobering and challenging. He began with a disarming admission:

> We are attempting to manage wildlife, but it is by no means certain that we shall succeed, or that this will be our most important contribution to the design for living. For example, we may, without knowing it, be helping to write a new definition of what science is for. We are not scientists. We disqualify ourselves at the outset by professing loyalty to and affection for a thing: wildlife. A scientist in the old sense may have no loyalties except to abstractions, no affections except for his own kind. (Leopold 1940, 343)

Through the 1940s and especially with the unfolding of events during and after World War II, Leopold became increasingly disenchanted with the course of the modern scientific enterprise. Although himself a pioneer in a new branch of basic and applied sciences, he saw the drift toward what he considered misapplied science as a grave danger for which scientists should not shirk responsibility. Already, by 1940, he was voicing concern over this trend. He told The Wildlife Society audience, "The definitions of science written by, let us say, the National Academy, deal almost exclusively with the creation and exercise of power. But what about the creation and exercise of wonder or respect for workmanship in nature?" (Leopold 1940, 343). To Leopold, this was not merely a peripheral concern for the new wave of "wildlifers"; it lay at the very core of their work. Mincing no words, he warned that "unless we can help rewrite the objectives of science, our job is predestined to failure" (Leopold 1940, 344).

Rewriting the "objectives of science" plainly took matters of wildlife conservation beyond the realm of science per se and into the realm of the arts and letters, ethics and philosophy. From his earliest days as a young hunter, aspiring ornithologist, and outdoor adventurer, Leopold tended to take an integrated approach to conservation matters (Meine 1988). He carried this tendency over into his professional life. As a forester, he helped to push the boundaries of the field outward to include issues of soil erosion, recreation, game protection, and wilderness preservation. In the heady days of the 1930s, as wildlife conservation was metamorphosing, he kept broad intellectual margins, regularly drawing connections to other disciplines in his writing. As one of the respected elders of the profession, having seen it through its infancy and preparing it for its adolescence, he reasserted the point:

> Our profession began with the job of producing something to shoot. However important this may seem to us, it is not important to the emancipated moderns who no longer feel soil between their toes. We find that we cannot produce much to shoot until the landowner changes his way of using land, and he in turn cannot change his ways until his teachers, bankers, customers, editors, governors and trespassers change their ideas about what land is for. To change ideas about what land is for is to change ideas about what anything is for. Thus we started to move a straw, and end up with the job of moving mountains. (Leopold 1940, 346)

How does one move a mountain? Not quickly, and not easily. To develop new wildlife management techniques—to document food habits, conduct life-history studies, develop census methods, understand how the pattern of cover types influences populations, and so forth—was the daily work of a rapidly growing number of students and researchers. To develop new modes of perception and a new philosophy of land use was the work of generations and of other domains of the human mind. Having defined the technical foundations of the field, Leopold now challenged his professional progeny to not neglect this more complex task: "I daresay few wildlife managers have any intent or desire to contribute to art and literature, yet the ecological dramas which we must discover if we are to manage wildlife are inferior only to the human drama as the subject matter for the fine arts" (Leopold 1940, 344). Even as wildlife ecology was beginning to grow confident in its expanded role as a science, its chief scientist was advising its adherents to surmount "the senseless barrier between science and art" (Leopold 1940, 344).

Had Leopold himself neglected the humanistic aspects of wildlife conservation, he would still be remembered as a highly effective thinker, leader, activist, and teacher. In his own view, however, the profession would have remained incomplete, its scope restricted. The necessary (and incomplete) task of integrating wildlife management into a still more comprehensive conservation vision would have been further postponed. Leopold understood that the humanities must help to define "the objectives of science." In articulating that need, Leopold established a very high standard for his colleagues. In meeting that standard with *A Sand County Almanac,* he provided an exemplary model for them.

Evolution of a Classic

The nuanced voice of Aldo Leopold that readers would come to know through *A Sand County Almanac* was late in developing. At the time of his 1940 presidential address to The Wildlife Society, Leopold was fifty-three years old. He had not yet even begun to think about the collection of essays that became the *Almanac.* That his thoughts had been moving in that general direction, however, can be seen in his published works at the time. Leopold's pen had always been sharp, productive, and provocative. His output of scientific papers, professional reports, editorials, policy statements, popular articles, and essays already amounted to some two hundred publications. In the late 1930s, however, he began to find new possibilities in his literary voice.

The first outward indication of this came in January 1937, with publication of the essay "The Thick-Billed Parrot in Chihuahua" in *The Condor* (see Table 1.1). The previous fall, Leopold and a friend had ventured into Mexico's Sierra Madre Occidental for a two-week bow-hunting trip. This trip (and another at the end of 1937) would have a lasting impact on Leopold's views of land management (Meine 1988, 367–368; Nabhan 1997). It also had an immediate impact on the tone of his public writing. "The Thick-Billed Parrot in Chihuahua" was an unusually evocative account for Leopold and no doubt for the readers of *The Condor.* In it, Leopold directly addressed the relationship between scientific understanding and the apprehension of beauty. Seeking to convey the "imponderable essence" of the Sierra Madre, he focused on the riotous calls of the "roistering" parrots. "As a proper ornithologist," he wrote, "I should doubtless try to describe the call. It superficially resembles that of the piñon jay, but the music of the pinoneros is soft and nostalgic as the haze hanging in their native canyons, while that of the Guacamaja is louder and full of the salty enthusiasm of high comedy" (Leopold 1937c, 10). Leopold later included the essay, in revised form, in *A Sand County Almanac* under the title "Guacamaja" (the native name for the thick-billed parrot). It was, in essence, the first installment in the collection of essays to come. (See table 1.1; the essay "The Alder Fork" and a portion of "The Land Ethic" had appeared in print in 1932 and 1933, respectively, but were only fitted into the evolving manuscript of *A Sand County Almanac* at a late stage in its development.)

Emboldened perhaps by *The Condor*'s publication of his rather unorthodox essay, Leopold followed it up with "Marshland Elegy," his haunting ode to cranes and their wetland homes (Leopold 1937a), published in *American Forests* in October of 1937. A few months later, *Bird-Lore,* the periodical of the National Audubon Society, published "Conservation Esthetic," in which Leopold rehearsed the point that he would make later in his presidential address to The Wildlife Society: "Let no man jump to the conclusion that [he] must take his Ph.D. in ecology before he can 'see' his country. On the contrary, the Ph.D. may become as callous as an undertaker to the mysteries at which he officiates" (Leopold 1938, 107). Leopold plainly appreciated the

TABLE 1.1. This chronology of events leading up to the publication of *A Sand County Almanac* draws on information in the Aldo Leopold Papers at the University of Wisconsin–Madison and in publications by Flader (1974), Ribbens (1987), and Meine (1988).

Jan.–Feb. 1937	*The Condor* publishes Leopold's "The Thick-Billed Parrot in Chihuahua."
Oct. 1937	*American Forests* publishes Leopold's "Marshland Elegy."
Mar.–Apr. 1938	*Bird–Lore* publishes Leopold's "Conservation Esthetic."
Nov. 1938	*Wisconsin Agriculturist and Farmer* publishes the first of Leopold's series of short essays on farm wildlife.
Early 1941	Leopold discusses possible collaboration with his friend and student H. Albert Hochbaum.
Nov. 1941–Jan. 1942	Alfred A. Knopf indicates interest in Leopold's collection of essays in initial exchange of correspondence.
Jan. 1943–Jun. 1944	Leopold drafts and revises many essays; he corresponds regularly with Hochbaum over the nature and narrative stance of the collection.
1943	*Wisconsin Conservation Bulletin* begins publishing short essays by Leopold.
8 Sep. 1943	Leopold drafts "Great Possessions."
Jan. 1944	Leopold works steadily on the essays; the working title for the collection is "Marshland Elegy—And Other Essays."
1 Apr. 1944	Leopold drafts "Thinking Like a Mountain."
Late Apr. 1944	Leopold meets Macmillan Company editor at the Ninth North American Wildlife Conference.
6 Jun. 1944	Leopold sends thirteen essays to Macmillan Company under the title "Thinking Like a Mountain—And Other Essays"; he sends the same essays to Knopf two days later.
July 1944	Rejection letter from Macmillan Company.
24 July 1944	Rejection letter from Knopf.
24 Aug. 1944	Knopf sends follow–up letter, with reviewers' comments, to Leopold.
Nov. 1944	Leopold writes to Hochbaum that he is "flirting with the almanac idea" for the essay collection.

(*continued*)

TABLE 1.1. (*continued*)

19 Jan. 1946	Leopold forwards essays to the University of Minnesota Press; they are rejected two weeks later.
Apr. 1946	Leopold's correspondence with Knopf resumes.
Fall 1946	Leopold undertakes extensive drafting and revising of essays.
31 Oct. 1946	Initial communication between Leopold and Oxford University Press.
Feb. 1947	Albert Hochbaum removes himself from the project.
July 1947	Leopold overhauls structure of the collection, adopts new title ("Great Possessions"), composes "The Land Ethic," and drafts foreword.
5 Sep. 1947	Leopold sends new manuscript to Knopf.
14 Sep. 1947	Leopold meets with Charles Schwartz in St. Louis to discuss the manuscript and illustrations; Schwartz agrees soon thereafter to provide illustrations.
Fall 1947–Winter 1948	Leopold continues to draft and revise essays.
5 Nov. 1947	Knopf rejects manuscript and suggests that the collection as a whole be recast.
Nov. 1947	Luna Leopold approaches Oxford University Press about publication of the collection; William Vogt simultaneously approaches William Sloane Associates.
5 Dec. 1947	Leopold revises foreword.
19 Dec. 1947	Leopold sends manuscript to Oxford University Press and to William Sloane Associates.
4 Mar. 1948	Leopold again revises foreword.
14 Apr. 1948	Leopold receives telephone call from Oxford University Press accepting his manuscript for publication; Oxford's acceptance letter is sent the same day.
21 Apr. 1948	Leopold dies while fighting fire on a neighbor's property.
22 Apr. 1948	Oxford writes to Leopold's student Joe Hickey expressing concern over the publication of the collection.
Apr. 1948–Dec. 1948	Luna Leopold and Joe Hickey, working with other Leopold family members and students, oversee final editing and preparation of manuscript; after extensive discussions, the title is changed to *A Sand County Almanac and Sketches Here and There.*

TABLE 1.1. (*continued*)

Fall 1949	*A Sand County Almanac* is published.
1953	Oxford University Press publishes *Round River: From the Journals of Aldo Leopold.*
1966	Oxford University Press reissues the *Almanac* in an enlarged edition as *A Sand County Almanac with Other Essays on Conservation from Round River.*
1968	Oxford University Press publishes original volume in paperback.
1970	Sierra Club/Ballantine Books publishes enlarged edition in paperback.

need for antidotes to insensitivity. Even as science was improving the capacity of wildlife conservationists to analyze and manage, Leopold suggested that aesthetic awareness would be needed to enhance their capacity to perceive (Callicott 1987). It was as if Leopold, recognizing that wildlife management's scientific underpinnings were finally well set, felt freer now to attend to its cultural and ethical bases.

In November 1938, Leopold produced the first in a series of concise essays on farm wildlife for the *Wisconsin Agriculturist and Farmer.* Over the next three to four years, twenty-nine of these informative essays would appear in the widely distributed periodical. (From 1943 to 1945, Leopold would publish a similar series in the *Wisconsin Conservation Bulletin.*) Several of these short pieces would later be incorporated, in revised form, into *A Sand County Almanac.* The most significant impact of the series, however, may have been to oblige Leopold to communicate regularly with a general audience. His growing experience as a college instructor during these years also seemed to have increased his dedication to this task of raising the general level of ecological literacy. "The citizen-conservationist," Leopold wrote in 1937, "needs an understanding of wildlife ecology not only to enable him to function as a critic of sound policy, but to enable him to derive maximum enjoyment from his contacts with the land" (Leopold 1937b, 80). Between 1939 and 1941, this intent was evident not only in Leopold's farm wildlife essays, but also in several other publications, including "Song of the Gavilan," "Escudilla," and "Cheat Takes Over" (first published in the *Journal of Wildlife Management, American Forests,* and *The Land,* respectively).

By the summer of 1941, Leopold had begun to think about bringing several of his essays together into a volume. In November of that year, an editor at Alfred A. Knopf wrote to Leopold indicating interest in "a good book on wildlife observation . . . a personal book recounting adventures in the field." As the editor saw it, this book should appeal to laymen while allowing the author the opportunity to offer "opinions on ecology and conservation" (Ribbens 1987). Leopold, it happened, had already been discussing such a project with his graduate student H. Albert Hochbaum, a talented artist and waterfowl biologist. Hochbaum and Leopold were both heavily burdened with their normal workload, but they agreed to work together as time allowed. Their intense, sometimes rocky, but mutually challenging collaboration over the next several years would prove critical to the ultimate tone of the collection as a whole (Ribbens 1987; Meine 1988, 460–461).

Leopold soon found himself with more time to devote to this project. As the United States entered World War II and students departed from the University of Wisconsin campus, Leopold's teaching and advising load ebbed. Leopold maintained a busy pen through 1942, but it was not until he received a follow-up inquiry from Knopf in April 1943 that he again focused his attentions on the proposed collection. Over the next year, Leopold drafted and redrafted some of his most memorable essays. Among these, importantly, were several that drew upon Leopold's activities with his family at the exhausted piece of farm property that he had acquired in 1935. These essays (in particular "Great Possessions") gave a more personal tone to the evolving book.

Albert Hochbaum, who was carefully reading and reviewing the drafted essays, recognized this as a turning point in Leopold's literary development. In one of many blunt but respectful exchanges between them during this period, Hochbaum encouraged Leopold in this new direction. "This series of sketches brings the man [Leopold] himself into focus. . . . As you round out this collection, take a sidewise glance at this fellow and decide just how much of him you want to put on paper" (Meine 1988, 457). Less than a month later, Leopold responded to Hochbaum's prodding with "Thinking Like a

Mountain," the famous account of his killing of a wolf during his youthful days as a forester in the American Southwest. Committed to the new course the collection was now taking, Leopold changed its working title to "Thinking Like a Mountain—And Other Essays."

By June 1944, Leopold's manuscript included thirteen essays (Ribbens 1987). He sent these off to Knopf and to an editor at the Macmillan Company who had also expressed interest in Leopold's writing. Both publishers turned down the manuscript. Macmillan, citing wartime paper shortages, rejected it outright. Knopf's editor felt that the essays were simply too varied in tone, length, and subject to hang together. The Knopf review, however, gave Leopold room for hope that, with extensive revision and additional essays, these stylistic and structural problems could be overcome. By the end of 1944, Leopold indicated to Hochbaum that he was playing with "the almanac idea . . . as a means of giving 'unity' to my scattered essays." The earlier farm wildlife essays seemed to have prompted him to consider the almanac format; in any case, this was the first mention of it in the context of the evolving collection (Ribbens 1987).

Other professional obligations absorbed Leopold's time over the following year. Not until the war was over, another rejection letter was received (this one from the University of Minnesota Press), and the correspondence with Knopf was reestablished did Leopold return to his disparate batch of essays. In correspondence with Knopf in the spring of 1946, Leopold suggested that he might add several of the more "philosophical" essays he had published in professional journals—thus making the unity of the essays even more problematic. Knopf's skeptical but supportive editor pointed out the difficulty in "fitting the pieces together in a way that will not seem haphazard or annoying to the reader" (Ribbens 1987).

Through the remainder of 1946 and into early 1947, this would remain a quandary for Leopold. Once more, other responsibilities (and a substantial influx of returning students) prevented him from focusing on his extracurricular writing. What little time he had to spare for the essays usually found him, before dawn, at his desk in his university office, wielding the pencils and yellow legal pads that he typically used in his later years. Leopold rarely wrote at the family's "shack" or elsewhere in the field, and his meticulous journals were filled not with literary expression, but with detailed phenological records, field observations, and other scientific data. Although Leopold was unable to work on his manuscript with any regularity during this time, he intermittently drafted new essays and revised older ones. And he continued to wrestle with the essential dilemma of the collection: how to meld into a coherent whole his descriptive field sketches, his ecological cautionary tales, and his statements of conservation philosophy.

As of spring 1947, the manuscript hung in limbo. Because of other commitments, Hochbaum had withdrawn as illustrator. Leopold, as the head and sole faculty member of his department, was preoccupied with accommodating the postwar boom in student enrollment. And he was increasingly distracted by the painful facial spasms associated with trigeminal neuralgia (or tic douloureux), with which he had been afflicted since late 1945.

Finally, in the summer of 1947, Leopold found time to devote himself exclusively to the essays. In this crucial period, the collection, which Leopold was now calling "Great Possessions," assumed the form that its eventual readers would recognize. Leopold drafted a lengthy foreword that provided autobiographical context for the essays. He divided the manuscript into three parts. In the first, he used the almanac format to bring order to the "shack" essays. In the second, he gathered the recollections and ecological interpretations of other landscapes in his experience. In the third section, he included four of his more conceptual discussions of conservation themes, including his newly synthesized summary statement, "The Land Ethic" (Meine 1987). With renewed hope, Leopold sent the overhauled manuscript back to Knopf in early September. Scheduled to undergo brain surgery later that month at the Mayo Clinic in Minnesota, Leopold had made his summer one of determined and uninterrupted concentration. The rejection letter from Knopf arrived in early November. Knopf's editors again found the collection "far from being satisfactorily organized as a book," adding that it was "unlikely to win approval from readers or to be a successful publication as it now stands" (Meine 1988, 509). Giving up on Knopf, Leopold allowed his son Luna to assume the role of literary agent. While Luna approached Oxford University Press, Leopold's close colleague, William Vogt, brought the manuscript to the attention of William Sloane Associates (who would soon publish Vogt's own conservation classic, *Road to Survival*).

Although disappointed and frustrated by Knopf's rejection, Leopold responded quickly. He rewrote the long foreword ("the better to orient the reader on how and why the essays add up to a single idea" [Leopold 1947]) and, in December 1947, sent the manuscript to the two new prospective publishers. The four previous rejections might have hobbled Leopold's expectations, but over the winter, he continued to draft new essays (including "Good Oak"). Following Luna's recommendation, he secured a new illustrator, Charles Schwartz, then working with the Missouri Conservation Commission.

As Leopold recuperated fitfully from his surgery, he waited for word. Both publishers were reading the manuscript with approval. Oxford responded first. On 14 April 1948, Oxford editor Philip Vaudrin called Leopold in Madison to inform him that the manuscript had been accepted for publication. They discussed plans for final revisions, with the goal of having the book available in the fall of 1949. One week later, on 21 April, Leopold suffered a fatal heart attack while fighting a neighbor's grass fire near the shack.

After the profound shock of Leopold's death had eased, Luna Leopold assumed responsibility for seeing the manuscript through to publication. Working with Joe Hickey, Frances and Frederick Hamerstrom, and other close colleagues and students of Leopold, Luna negotiated the final terms of publication with Oxford University Press. This team collaborated in making final editorial decisions. Several essays were added, shifted, or renamed, but most of the alterations to Leopold's manuscript were minor. The team felt that it was better to leave Leopold's work intact than to risk making inappropriate changes.

Luna Leopold did agree, reluctantly, to one significant change. Oxford considered Leopold's title, "Great Possessions," too obscure and too Dickensian. Consultations among Oxford's editors, Luna, and the editorial panel yielded several alternative titles, none of which seemed to capture the book's characteristic tone of concern tempered by gentle irony and understated wonder. In the end, they chose for the title the heading of the manuscript's first section, "A Sand County Almanac." Oxford published the book in the fall of 1949 under the full title *A Sand County Almanac and Sketches Here and There* (Meine 1988, 523–524).

Humanizing Conservation

This condensed narrative cannot fully convey the impact of contemporary events, professional experiences, and private interactions on Leopold's evolving vision for his book (and, more broadly, for the conservation movement). However, even this rudimentary account reveals that Leopold was deeply devoted to the project's overarching goal—so much so that he persisted through multiple rejections, continual questioning of content and style, and a series of difficult personal challenges. The goal was to break down "the senseless barrier between science and art," to unite informed observation of the living world, through the lens of ecology and evolutionary biology, with an enriched appreciation of the world's inherent beauty and drama. "There is no other way," he wrote in the final foreword, "for land to survive the impact of mechanized man, nor for us to reap from it the esthetic harvest it is capable, under science, of contributing to culture" (Leopold 1949, viii).

At the same time, Leopold plainly understood that this was not simply an exercise in ecological aesthetics. Throughout the 1940s, trends in world events, human relations, and human interactions with the natural world weighed heavily on Leopold and on many of his like-minded colleagues in the conservation movement. A careful reading of *A Sand County Almanac* provides ample clues that this was definitely a book of its time. From "Pines above the Snow": "[T]he 1941 growth was long in all pines; perhaps they saw the shadow of things to come, and made a special effort to show the world that pines still know where they are going, even though men do not" (Leopold 1949, 83). From "Wilderness": "Ability to see the cultural value of wilderness boils down, in the last analysis, to a question of intellectual humility. The shallow-minded modern who has lost his rootage in the land assumes that he has already discovered what is important; it is such who prate of empires, political or economic, that will last a thousand years" (Leopold 1949, 200). From "The Land Ethic": "In human history, we have learned (I hope) that the conqueror role is eventually self-defeating" (Leopold 1949, 204). The value of this new way of looking at the natural world rested in its capacity not only to enhance human awareness and appreciation, but also to improve our chances of achieving "harmony with land" (Leopold 1953,

155)—Leopold's definition of conservation. (And it might also have something to offer in our efforts to achieve more decent human relations.)

Those chances seemed to be diminishing at the time. Faced with the postwar prospect of unprecedented economic and technological changes, and overwhelmed by the shift away from the field-oriented biology at which he excelled, Leopold spared no words in his critique of the forces driving the scientific agenda. In a 1946 address to the Wisconsin Society for Ornithology, he stated, "Science, as now decanted for public consumption, is mainly a race for power. Science has no respect for the land as a community of organisms, no concept of man as a fellow passenger in the odyssey of evolution" (Meine 1988, 483).

He was equally forthright in criticizing the various component fields of conservation to which he himself had contributed so importantly. He shared with his students his concern that conservation, too, suffered from the fallacy, "clearly borrowed from modern science, that the human relation to land is only economic. It is, or should be, esthetic as well. In this respect our current culture, and especially our science, is false, ignoble, and self-destructive" (Flader and Callicott 1991, 337). Harsh words to cast upon the ears of listening undergraduates. Ecology, he would state in another context, opens one's senses to a "world of wounds" (Leopold 1953, 165).

Characteristically, Leopold lightened his message to his students by pointing out the fringe benefits of ecological literacy: "I am trying to teach you that this alphabet of 'natural objects' spells out a story, which he who runs may read—if he knows how. Once you learn to read the land, I have no fear of what you will do to it, or with it. And I know many pleasant things it will do to you" (Flader and Callicott 1991, 337).

Through *A Sand County Almanac,* Leopold sought to teach others to read the land, to recognize the wounds, and to savor the pleasures. By his very tone, he conveyed his trust in their ability to do so, and to act on what they read, learned, and enjoyed. This was for Leopold the solid foundation upon which conservation had to be built. In his unassuming and idiosyncratic book of essays, Leopold showed that we may move mountains by allowing the mountains—and the skies, the oceans, the freshwaters, the marshes, the forests, the prairies, the deserts, and all of the lives, human and otherwise, they contain—to move us.

References

Callicott, J. B. 1987. The land aesthetic. In *Companion to* A Sand County Almanac: *interpretive and critical essays*, ed. J. B. Callicott, 157–171. Madison: University of Wisconsin Press.

Dunlap, T. L. 1988. *Saving America's wildlife*. Princeton, N.J.: Princeton University Press.

Flader, S. L. 1974. *Thinking like a mountain: Aldo Leopold and the evolution of an ecological attitude toward deer, wolves, and forests*. Columbia: University of Missouri Press.

Flader, S. L., and J. B. Callicott, eds. 1991. The River of the Mother of God *and other essays by Aldo Leopold*. Madison: University of Wisconsin Press.

Leopold, A. 1930. Report to the American Game Conference on an American game policy. In *Transactions of the Seventeenth Annual American Game Conference*. New York: American Game Association.

Leopold, A. 1931. *Report on a game survey of the north central states*. Madison, Wisc.: The Democrat Press, for the Sporting Arms and Ammunitions Manufacturers' Institute.

Leopold, A. 1933. *Game management*. New York: Charles Scribner's Sons.

Leopold, A. 1936. Threatened species. *American Forests* 42:116–119.

Leopold, A. 1937a. Marshland elegy. *American Forests* 43:472–474.

Leopold, A. 1937b. Teaching wildlife conservation in public schools. *Transactions of the Wisconsin Academy of Sciences, Arts, and Letters* 30:77–86.

Leopold, A. 1937c. The thick-billed parrot in Chihuahua. *Condor* 39:9–10.

Leopold, A. 1938. Conservation esthetic. *Bird-Lore* 40:101–109.

Leopold, A. 1939. A biotic view of land. *Journal of Forestry* 37:727–730.

Leopold, A. 1940. The state of the profession. *Journal of Wildlife Management* 4:343–346.

Leopold, A. 1947. Letter to Philip Vandren, 28 November.

Leopold, A. 1949. *A sand county almanac and sketches here and there*. New York: Oxford University Press.

Leopold, A. 1953. *Round River: from the journals of Aldo Leopold*. L. B. Leopold, ed. New York: Oxford University Press.

McCabe, R. E., ed. 1988. *Aldo Leopold: mentor*. Department of Wildlife Ecology, University of Wisconsin–Madison.

Meine, C. D. 1987. Building "The Land Ethic." In *Companion to* A Sand County Almanac: *interpretive and critical essays*, ed. J. B. Callicott, 172–185. Madison: University of Wisconsin Press.

Meine, C. D. 1988. *Aldo Leopold: his life and work*. Madison: University of Wisconsin Press.

Meine, C. D. 1994. The oldest task in human history. In *A new century for natural resources management*, eds. R. L. Knight and S. F. Bates, 7–35. Washington, D.C.: Island Press.

Nabhan, G. P. 1997. Sierra Madre upshot: ecological and agricultural health. In *Cultures of habitat: on nature, culture, and story*, 43–56. Washington, D.C.: Counterpoint Press.

Ribbens, D. 1987. The making of *A Sand County Almanac.* In *Companion to* A Sand County Almanac: *interpretive and critical essays*, ed. J. B. Callicott, 91–109. Madison: University of Wisconsin Press.

Trefethen, J. B. 1975. *An American crusade for wildlife.* New York: Winchester Press and the Boone and Crockett Club.

A Sense of Place, A Sense of Time

Chapter 2

Aldo Leopold

Blending Conversations about Public and Private Lands

RICHARD L. KNIGHT

Among the shelves of published American works, there resides a slim volume of essays written by an Iowan who became a forester, then a wildlife manager, and later a teacher, and who died a visionary. These essays, written during the early morning hours in his office on the Madison campus of the University of Wisconsin, were drawn from his experiences at a shack on a sand county farm along the banks of the Wisconsin River. Aldo Leopold's writings in *A Sand County Almanac* touch the part of Americans that E. O. Wilson had in mind when he coined the word "biophilia" (Wilson 1984). Because we evolved from nature, we still carry a part of nature in our hearts, in our souls, and in our heads. This is where humans feel their relationship with and responsibilities to the land. Few have equaled Leopold's ability to evoke the complexity of human-land relations. And, central to this essay, none has written such balanced words regarding our obligations to public and private lands.

Public and Private Lands

The time is more urgent than ever for honest conversations about public and private lands because we stand at a threshold. Do we continue our preoccupation with viewing land as property and argue over our rights to do with it as we wish, or do we engage in conversations that focus on land as the wellspring of human health, happiness, and prosperity (Baskin 1997)? If we so vociferously defend our property rights, where do we include in our discussions of land our responsibilities to it? These questions are relevant and rise out of a dichotomy in American society. On the one hand, Americans appear to be increasingly self-absorbed, more interested in how they will benefit rather than what they will bestow. On the other, Americans are becoming more involved in land-related issues, reflected in the phenomenal growth in open-space initiatives and land trusts and in the strongly pro-environment sentiments expressed in public opinion surveys.

Critical to these discussions about land are conversations involving the milieus of public and private lands. Although our lands are dissected by administrative boundaries of federal, state, and local jurisdictions, and further fragmented by private ownership, they are part of a much greater ecological fabric defined by watersheds, flyways, plant communities, ecological processes such as fire, and other powerful natural phenomena that almost always defy legal boundaries. Our discussions about what type of society we are building will not be complete until they involve both public and private lands, for they are entwined ecologically, politically, culturally, and economically (Knight and Landres 1998).

Aldo Leopold was aware of the distinctions and commonalities between public and private lands. He understood how human-placed, arbitrary boundary lines had divided the landscape, but he also believed that humans had the power to blend artificially fragmented landscapes back together again. Leopold was well prepared for such a holistic approach to land stewardship. He had spent half of his career attempting to understand the public lands of the American Southwest. Later, he owned a sand county farm where he studied the flow of nutrients, weeds, and wildlife that crossed the administrative lines separating him from his neighbors. These experiences led Leopold to believe that lines could be dimmed across these boundaries: "It is a fact, patent both to my dog and myself, that at daybreak I am the sole owner of all the acres I can walk over. It is not only boundaries that disappear but also the thought of being bounded" (Leopold 1966, 41).

Leopold struggled with the dualism of public and private lands; he understood that it was more than simply boundaries and survey lines that made them separate to Americans. We view and value these lands in quite different ways. Public lands belong to all Americans; they are part of everyone's inheritance. Private lands, however, belong to only a few with their special privileges bequeathed by the Constitution and Bill of Rights. Leopold appreciated that we had rights to land, both public and private, but he also regretted that we felt so few responsibilities to these same lands.

There have always been tensions among Americans concerning their rights and responsibilities to land. For much of America's early history, the government was primarily interested in giving land away. Then, with what was left—the public lands—the government became increasingly involved with trying to determine the right blend of uses that should be allowed (Wilkinson 1992). Indeed, the public land base has never been static, and it was greatly augmented during the dry years of the 1930s, when much private land was given back to the government by default. The land's value had been used up by landowners, and neither banks nor county or state governments wanted the land. Today, there is considerable heat and rhetoric among people who espouse so-called private property rights and who lash out against the government for infringing on their liberties to do as they like with both public and private lands (Echeverria and Eby 1995). If these individuals had studied their American history better, they would understand how similar, self-serving actions of earlier generations had robbed the land of much of its ecological and economic value. In the end, with few exceptions, only our government has been able to show any consistent concern for land health (Stegner 1987).

But there are encouraging signs today in America regarding public and private lands. Public land management has evolved from the concept of multiple use to ecosystem management (Grumbine 1992). Although not very different in principle, these two approaches are radically different in practice. Under multiple use, we overused our lands. Under ecosystem management, we have another chance to rediscover our stewardship responsibilities. Stewardship is defined by our capacity to place land health above the land's many uses. Ecosystem management is not unlike multiple use in that it recognizes commodity and amenity uses of our lands, but it is quite different from multiple use in that it places the protection of native biological diversity on an equal footing with these customary uses. In addition, ecosystem management promotes ecological restoration, stresses the importance of the human dimension, and encourages land managers to forsake their "command-and-control" mentality and replace it with an outlook that encourages cooperation. It rewards managers for crossing administrative lines and finding new ways to work cooperatively with public and private constituencies (Knight and Meffe 1997). Whether ecosystem management will actually allow us to place land health above land use is the great experiment now under way. We have never before been able to stick to this slippery and difficult path, always succumbing to an addiction to dollars before land health.

Discussions about private land are also evolving for the better (Weeks 1996; Beatley and Manning 1997). We Americans have always viewed land as property. Indeed, in our Constitution, the word *land* does not occur but the word *property* does. Because we view private ownership of land under the rubric of property, we tend to emphasize our rights rather than our responsibilities. With every passing year, however, there are more exceptions to this generalization. From coast to coast, citizens are forming alliances to promote land health and engage in stewardship projects. My neighbors are Catherine and Evan Roberts. They ranch in Livermore, Colorado, and are what's left of a fifth-generation ranch family. They are trying to ensure that their land stays in agriculture and out of residential development, increasingly the fate of private lands in America (Knight 1997). When asked how they view their ranch, they reply, "We have never felt like we owned the land, we have always felt like it owned us." Catherine and Evan believe that there are two kinds of Americans today, the "takers" and the "caretakers." They have always felt like they belonged to the latter, placing their obligations to the land on a higher level than their rights to exploit it.

Who Speaks for the Land?

In conversations about public and private lands, it is appropriate to ask, Who speaks for the land? Who are its advocates? Who sees the land as an entity not only to own but also to belong to? Who recognizes that along with ownership of land comes responsibility to both the human and natural communities? Who will champion what Aldo Leopold proclaimed:

> That there is some basic fallacy in present-day conservation is shown by our response to it. Instead of living it, we turn it over to bureaus. Even the

> landowner, who has the best opportunity to practice it, thinks of it as something for government to worry about. I think I know what the fallacy is. It is the assumption, clearly borrowed from modern science, that the human relation to land is only economic. It is, or should be, esthetic as well. In this respect our present culture . . . is false, ignoble, and self-destructive. (Flader and Callicott 1991, 337)

I would guess that when most Americans think about land outside the city, they envision national parks and forests—the gems that dazzle in our nation's crown: Yosemite, Mesa Verde, Acadia, Mount Rainier, Yellowstone, Great Smoky Mountains. We wait in lines at their entrance stations; we flock to their visitor centers. But, as conservationists, we must also consider our private lands—the working lands, or better put, the "middle lands" of America. Who will speak for these landscapes that constitute most of our country's land base? Who will attend to the lands situated between our cities on the one hand and our parks and wilderness areas on the other, the lands that historically have served as both the breadbasket and the heartland of our nation?

Biodiversity, ecosystem management, sustainable development, and ecological restoration are concepts as relevant to our nation's private lands as they, seemingly overnight, have become to our public lands. And, importantly, as we begin a new century and enter a new millennium, it appears that history and public policy have converged to forever shift America's preoccupation with public lands to a larger commitment to stewardship for all lands, public and private. More than at any time in decades, Americans seem ready for discourse on both public and private land issues. We are ready to capture the true spirit of integrated land conservation that Leopold pegged:

> I do not challenge the purchase of public lands for conservation. . . . I do challenge the growing assumption that bigger buying is a substitute for private conservation practice. Bigger buying, I fear, is serving as an escape-mechanism—it masks our failure to solve the harder problem. The geographic cards are stacked against its ultimate success. It is exactly as effective as buying half an umbrella. (Flader and Callicott 1991, 196–197)

Leopold's Voice

Aldo Leopold's writings reflect a balanced consideration and concern for public and private land relationships. Raised in Burlington, Iowa, he knew only private land in his youth (Meine 1988). He prowled the river sloughs and bottomland forests of the Mississippi. He walked the rich mosaic of farmland, prairie groves, and marshes. He counted the roosting flights of wood ducks on fall evenings and hunted prairie chickens on remnant tall-grass plains. However, when age, education, and a job allowed, he migrated to the American West and its immense public domain.

For fifteen years he worked the national forests of Arizona and New Mexico. And he learned a good deal. He began to see that human uses of land could be beneficial or harmful, depending on the user's sense of stewardship; that soil had to be carefully cultivated and was no more deeded to the land than the wind that brushed across it; that the behavior of deer, mountain lions, wolves, and grizzly bears was as intricately designed as the bones, sinews, and fibers that made their bodies so graceful; that humans could easily disrupt such intricate relationships. He became imbued with the mystery found in land still so vast and roadless that it could only be crossed on horseback or on foot. Out of Leopold's experiences came the spark that ignited the American wilderness movement, as important in its own way as the earlier idea of national parks, both of which indelibly mark America and Americans as lovers of public lands.

Leopold thought long and hard about what he had seen and learned on these public lands, and many of the essays in *A Sand County Almanac* reflect his concern for our relationship to these lands. His extended stay in the wildlands of the American Southwest influenced him profoundly. He wrote in *A Sand County Almanac*:

> A deep chesty bawl echoes from rimrock to rimrock, rolls down the mountain, and fades into the far blackness of the night. It is an out-burst of wild defiant sorrow, and of contempt for all the adversities of the world. Every living thing pays heed to that call. To the deer it is a reminder of the way of all flesh, to the pine a forecast of midnight scuffles and of blood upon the snow, to the coyotes a promise of gleanings to come, to the cowman a threat of red ink at the bank, to the hunter a challenge of

> fang against bullet. Yet behind these obvious and immediate hopes and fears there lies a deeper meaning, known only to the mountain itself. Only the mountain has lived long enough to listen objectively to the howl of a wolf. (Leopold 1966, 129)

Leopold's deep appreciation of wilderness, expressed in deeds as well as words, defined the wilderness movement. This would have been an achievement enough for one lifetime. But Leopold's thinking continued to mature and, before he was through, he left us the land ethic, a philosophy encapsulated in this statement: "We abuse land because we regard it as a commodity belonging to us. When we see land as a community to which we belong, we may begin to use it with love and respect" (Leopold 1966, x).

We might ask, If Leopold had stayed on the public lands of the American West, would he have ever written his land ethic? Could he have written something like this: "We end, I think, at what might be called the standard paradox of the twentieth century: our tools are better than we are, and grow better faster than we do. They suffice to crack the atom, to command the tides. But they do not suffice for the oldest task in human history: to live on a piece of land without spoiling it" (Flader and Callicott 1991, 254).

To live on a piece of land without spoiling it. One does not live on public lands. What life experiences, beyond his tenure on the public lands, led Leopold to plumb the depths of human relationships to land so fully that his writings ring true a half-century later? What experiences caused him to write of conservation increasingly as a protest against destructive land use?

> Government, no matter how good, can only do certain things. Government can't raise crops, maintain small scattered structures, administer small scattered areas, or bring to bear on small local matters that combination of solicitude, foresight, and skill which we call husbandry. Husbandry watches no clock, knows no season of cessation, and for the most part is paid in love, not dollars. . . . Husbandry is the heart of conservation. (Flader and Callicott 1991, 298)

What happened to Leopold, of course, is that he changed home ranges. Inexplicably, or so it seems in reading his biography, Leopold awoke one morning and looked eastward. He then moved back to the Midwest. Why? Why would he, with his New Mexican wife, Estella, leave the "land of enchantment" for Wisconsin, the "dairy-land state"? Why would he decide to shift his attention from a region defined by its abundance of public lands to an area of mostly privately owned farms?

We may never know the answer, but, to me, Leopold's move serves as a powerful metaphor for what is occurring across America today. Perhaps symbolically, Leopold's physical and intellectual shift from public to private lands heralded a major shift in thinking that today we see rippling across our land; it's a shift that will reright the scales and provide a new balance in discussions of public and private lands (Diamond and Noonan 1996).

Typically ahead of his time in thinking about important conservation issues, might Aldo Leopold also have anticipated the need to consider our private lands, our working landscapes? He did. His writings leave a rich legacy in this regard: "The crux of the problem is that every landowner is the custodian of two interests, not always identical, the public interest and his own. What we need is a private inducement or reward for the landowner who respects both interests in his actual land practice. All conservation problems—erosion, forestry, game, wild flowers, landscapes—ultimately boil down to this" (Leopold 1934).

So in discussing public and private land issues, I would suggest these conversations are forever entwined, indivisible on the land, even with our helter-skelter administrative, survey, and ownership boundaries. Even with our jumbled and diverse human land uses. Even with our determination, usually fueled by an economic motive, to find a use for every inch of land. I believe this because our public and private lands are America. The landscape that defines our homes is a mosaic of public and private lands. Leopold might never have developed his land ethic had he not moved back to the Midwest and studied conservation on agricultural lands. I believe that his land ethic—a philosophy that increasingly shapes our thinking in discussions of human-land relationships—required an equal measure of experiences on public and private lands before it was properly seasoned.

The Upshot

Leopold's land ethic, and the critical role private land played in its development, provides a useful bridge to the present. Today, the tides of history and the forces that define our society forecast changes that relate to both public and private lands. The importance of agricultural land in our nation's consciousness, while always present, has, nonetheless, receded during the past several decades (Berry 1981). As the American landscape became increasingly urban, Americans' preoccupation with the public lands as our playgrounds grew. Now, the tide has turned, and with changing demographics and sentiments, the fate of our private lands is once more coming to the foreground in our thinking.

Today, Americans are increasingly interested in the relationship among private land, its productivity, and its stewardship. Stewardship values, as measured in soil retention, water conservation, and biodiversity, are applied to working landscapes, where food and fiber are produced. This is the sort of agriculture that Leopold had in mind when he wrote, "Conservation means harmony between men and land. When land does well for its owner, and the owner does well by his land; when both end up better by reason of their partnership, we have conservation. When one or the other grows poorer, we do not" (Flader and Callicott 1991, 255).

The upshot of these changes in sentiment is that we are entering an era when Americans will become conscientious stewards of not only our national parks and wildlands, but also our middle grounds, the working landscapes that provide essential commodities. These lands are the embodiment of a rural America that seems to be disappearing and whose loss we miss more with the demise of every family farm and ranch (Beatley and Manning 1997). The Natural Resources Conservation Service (NRCS) epitomizes these changes from the federal perspective. Under the banner "America's private lands: a geography of hope," the NRCS is championing a movement that values stewardship activities on private land and reflects a shared responsibility between public and private interests (NRCS 1996). In the private sector, The Nature Conservancy is perhaps the most conspicuous organization nationwide that is practicing community-based conservation in efforts to blend use and conservation across ecoregions (Weeks 1996).

Americans are beginning to realize the importance of agricultural lands to their own personal happiness. We can no longer take for granted that these working landscapes will always remain intact or that they will always serve as the pleasing interface we drive through when going from town to city, or city to wilderness. In Colorado, over 270,000 acres of farm and ranch lands are converted annually to something else: housing developments, highways, industrial parks, and shopping malls. For this reason alone, Americans will increasingly turn to organizations and agencies who are involved in conversations about the two-thirds of our country that is private land.

It is the rapid loss of private lands to development that may lead Americans to finally champion the conservation of all lands, the soil and water that they protect, the multiple and sustainable uses of public and private lands, and our natural heritage, the only true birthright of all Americans. I sense that we are close to this point. One can see it in a hundred watershed and community-based conservation initiatives across America (Yaffee et al. 1996). Representatives of public and private lands and local and national organizations are forging new methods of doing business that emphasize cooperation, not conflict; community values, not individualism; land health, not land wealth; and land responsibilities, not land rights.

This was the conversation that Leopold had in mind. He wrote many evocative and suggestive words about the need for improved human-land relationships. By caring about both people and healthy landscapes, in *A Sand County Almanac* he left us a key to the wisdom we seek on how to live better, fuller lives. Let Leopold have the last word:

> The song of a river ordinarily means the tune that waters play on rock, root, and rapid. This song of the waters is audible to every ear, but there is other music in these hills, by no means audible to all. To hear a few notes of it you must first live here for a long time, and you must know the speech of the hills and rivers. Then on a still night, when the campfire is low and the Pleiades have climbed over the rimrocks, sit quietly and listen for a wolf to howl and think hard of everything you have seen and tried to understand. Then you may hear it—a vast pulsing harmony—its score inscribed on a thousand hills, its notes the lives and deaths of plants and animals, its rhythms spanning the seconds and centuries. (Leopold 1966, 149)

References

Baskin, Y. 1997. *The work of nature: how the diversity of life sustains us*. Washington, D.C.: Island Press.

Beatley, T., and K. Manning. 1997. *The ecology of place: planning for environment, economy, and community*. Washington, D.C.: Island Press.

Berry, W. 1981. *The gift of good land.* New York: North Point Press.
Diamond, H. L., and P. F. Noonan. 1996. *Land use in America.* Washington, D.C.: Island Press.
Echeverria, J., and R. B. Eby, eds. 1995. *Let the people judge: wise use and the private property rights movement.* Washington, D.C.: Island Press.
Flader, S. L., and J. B. Callicott, eds. 1991. The River of the Mother of God *and other essays by Aldo Leopold.* Madison: University of Wisconsin Press.
Grumbine, R. E. 1992. *Ghost bears: exploring the biodiversity crisis.* Washington, D.C.: Island Press.
Knight, R. L. 1997. Field report from the new American West. In *Wallace Stegner and the continental vision,* ed. C. Meine, 181–200. Washington, D.C.: Island Press.
Knight, R. L., and P. L. Landres, eds. 1998. *Stewardship across boundaries.* Washington, D.C.: Island Press.
Knight, R. L., and G. K. Meffe. 1997. Ecosystem management: agency liberation from command and control. *Wildlife Society Bulletin* 25:676–678.
Leopold, A. 1934. Some thoughts on recreational planning. *Parks and Recreation* 18:136–137.
Leopold, A. 1966. *A sand county almanac with other essays on conservation from Round River.* New York: Oxford University Press.
Meine, C. 1988. *Aldo Leopold: his life and work.* Madison: University of Wisconsin Press.
Natural Resources Conservation Service. 1996. *America's private land: a geography of hope.* Washington, D.C.: Natural Resources Conservation Service.
Stegner, W. 1987. *The American West as living space.* Ann Arbor: University of Michigan Press.
Weeks, W. 1996. *Beyond the ark: tools for an ecosystem approach to conservation.* Washington, D.C.: Island Press.
Wilkinson, C. F. 1992. *Crossing the next meridian: land, water, and the future of the West.* Washington, D.C.: Island Press.
Wilson, E. O. 1984. *Biophilia.* Cambridge, Mass.: Harvard University Press.
Yaffee, A. F., A. F. Philips, R. C. Frenty, P. W. Hardy, S. M. Maleki, and B. E. Thorpe. 1996. *Ecosystem management in the United States: an assessment of current experience.* Washington, D.C.: Island Press

Chapter 3

Aldo Leopold's Wilderness, Sand County, and My Garden

JOHN SEIDENSTICKER

The Palouse was withering hot in July 1969. Heat radiated from scorched, rolling fields, sunup to sundown. Images of cool meadows and clear, rapid creeks invaded my thoughts. I was to begin radio-tracking cougars in the Idaho Primitive Area in January. I needed to get into "the country" to explore limits and define expectations for my new equipment. But German barred my way. I had to pass a German exam, an archaic hurdle on the way to a Ph.D. Day in and day out I struggled with the German language. Despite my surname, I was not a particularly strong student. My increasing cognitive dissonance demanded relief. My solution: thirty minutes of Aldo Leopold's *A Sand County Almanac* after lunch each day as a treat. Al Franzmann, my fellow graduate student and a far more talented German linguist, advised that this was no solution; I was creating a mirage equal to any created by the scorching sun on the Palouse landscape, he told me. Learning German, not basking in Aldo Leopold's vision, was the point. I would wither with frustration.

In those half-hour interludes, I escaped, transported by Leopold's spare, beautiful prose and his vision of why and how we can relate to land. Just a little Leopold vision each day, then back to the engineering side of my schooling, learning German. Days passed, after-lunch snippets added up, and Leopold's vision began to be my own. *A Sand County Almanac* ended too quickly. That grass fire in the sand counties that cut Leopold's life short also cut his narrative short, I felt. I wanted him around to help flesh out his vision. Now I believe that this is the purity and beauty of Leopold's human-land vision: it is as fleshed out as it needs to be. The rest is up to us.

Living Wild

German out of the way, I immersed myself in following the day-to-day activities of cougars living in the large, natural landscape of what is known today as the Frank Church River of No Return Wilderness. The climate and rugged relief of the Salmon and Snake River canyons shaped the moral and social nature of all of us who lived there, as it shaped the social organization of the cats I was following. Leopold never visited the Salmon River Mountains, but his ideas had molded the character of the landscape that I was experiencing. Alarmed by the rapid loss of roadless mountain country in the southwestern United States in the early 1900s, he lobbied for the withdrawal from production of the last few large national forest roadless areas (Leopold 1947), like the deep canyons and mountain basins where I tracked my cougars. The Forest Service, in cooperation with the Idaho legislature, designated the Idaho Primitive Area in 1931. Road building and extractive activities such as logging and livestock grazing were banned here, but not hunting or other recreational activities deemed compatible with wilderness values.

Seeking to define wilderness, Leopold wrote of being present as a wolf died and watching "the fierce green fire dying in her eyes" (Leopold 1949, 130). And he wrote of how the last grizzly in the Southwest walked into the string of a set gun and shot itself. "Man always kills the things he loves, and so we the pioneers have killed our wilderness," he wrote (Leopold 1949, 148). How fortunate I am, I felt, living here and learning how a large mammalian predator-prey system functions in this wild area. I embraced and thrived in this country. Every day, I thanked Leopold and his contemporaries for their vision and actions that made my life here possible, for the fragments of wilderness that were left to be experienced. Unlike Leopold's experience of watching wild eyes dying, here was a place where I could devote my energy to trying to gaze into the eyes of a living cougar.

Public debate was inevitable in 1931 when more than 1,200,000 acres of Idaho's public land was designated a primitive area and withdrawn from most extractive uses. This debate was still going on when I was there, four decades later. Some people considered the remote canyons of the Idaho Primitive Area useless, bordering on dangerous, even noxious. They called it backcountry. This was the watershed of the River of No Return after all. For romantic preservationists in the tradition of John Muir, this primitive area designation secured a large area in a wild and pristine state, untainted by considerations of material and economic gain. For others, this designation simply represented a waste of nature, a lockup of resources that should remain open, there for the taking, available to anyone. The progressive conservation movement, which spawned our national forests, preached the gospel of efficiency and focused on multiple uses (Hays 1959).

All of us who lived in that country, that wildland, depended daily on the elk and mule deer we killed in the autumn for our winter larder, and on the cutthroat trout in summer. The first smoked steelhead of the spring was a reason to gather over drinks and reflect on how good life was there. Locally cut firewood was essential for heating our cabins and tents. Guiding hunters and fishermen was a major source of income. We needed some limited local grazing for pack stock. We revered the roadless part of wild and primitive, but we needed chainsaws for cutting firewood and light airplanes for transport and a connection to the outside. These things were necessary to support our life, or so we felt, and, thus, it was our right to have and use them.

I come from a ranching family, and, as a graduate student, I was trained in the resource-based conservation tradition that governed the Forest Service. Yet the vision of a land ethic that Leopold had given us resonated deep within me. However, I did not see how his land ethic offered a clear resolution, or even a reduction in the tension, between the value-driven ideologies shaping political debate on the future of the Idaho Primitive Area. Parts of the Idaho Primitive Area were being considered for inclusion into the National Wilderness and Wild River systems. Even as large and wild as this land was and even with the challenges I overcame each day just to live there, both my wilderness lifestyle and this wildland seemed vulnerable rather than robustly wild. Leopold had warned us how the wild could easily slip away. Reading the unfolding debate in the local and statewide newspapers, I too came to sense a profound unease that the wild would be lost to me in the rough-and-tumble of public land politics.

I could not shake the nagging feeling that the prospects for this wild area, or great portions of it, were shaky at best. People had radically different reactions to the concept of wild. Certainly our privileged way of life, a life subsidized by local natural resources, would change in changing our legal designation from a primitive area to a wilderness area—that was certain. My cougars, the cats I had gotten to know intimately, were protected by a five-year moratorium on hunting while we conducted this study; they would soon become game again. I attended a hearing on setting the boundaries of the new wilderness area and found myself in lockstep with an outfitter who lived and worked on Big Creek. We didn't agree on most things in life, but we agreed on two that night: make the wilderness area as big as possible and keep the trail bikes out. I continued to look to Leopold's land ethic for guidance, as a religious person would look to the Bible, but at that time I couldn't find within *A Sand County Almanac* a resolution to the on-the-ground conflict that was threatening my wildland.

I looked to the transcendentalists for some grounding: "in wildness is the preservation of the world" (Thoreau in Nash 1973, 84). I was living in wildness. After a day of radio-tracking and snow-tracking cougars, I would read Thoreau's *Walden* (Thoreau 1947) in the glow of a gas lamp in a winter camp. I had saved *Walden* for such a time and place. Thoreau's romantic-transcendental conservation ethic didn't resonate with me there, and I couldn't finish the book. Thoreau's Walden Pond and his wild were just down the railroad tracks from his mother's kitchen in a rural, human-dominated, northeastern landscape (Foster 1999). That seemed very tame to me at that time, compared to the wild experience I was living.

I reread Leopold in an almost religious fervor in an effort to calm the cognitive dissonance created by my background and training in utilitarian conservation and my day-to-day experience in a vulnerable wildland. I reread his words:

> Conservation is a state of harmony between men and land . . . a system of conservation based solely on economic self-interest is hopelessly lopsided. It tends to ignore, and thus eventually eliminate, many elements in the land community that lack commercial value but that are (as far as we

> know) essential to its healthy functioning. . . . It assumes falsely, I think, that the economic parts of the biotic clock will function without the uneconomic parts." (Leopold 1949, 207–214)

Leopold, I then realized, had encapsulated and included these competing value systems. He had constructed an overarching vision with his land ethic. I had been blocked in my view of the wild because I was seeking resolution in my mind in a limited dimension based on a single resource. But Leopold had clearly escaped this two-dimensional space and had soared into a three-dimensional world in his land ethic vision.

With my colleague, John Messick, I spent perhaps the most physically demanding day of my life trying to find a female cougar on her summer home area, in terrain incised with cliffs and stands of Douglas fir on sharp, protected slopes rising three thousand feet above the south bank of Big Creek. We had little water and no food that day. I wasn't sure I would survive until late that night when I finally dragged into camp. Perhaps I wouldn't have without John. "Why was I doing this?" I asked myself. I could easily die here.

I wouldn't have traded it for any other experience. I did it because it was what I valued. My dissonance and the public tension over this wilderness and the concept of wild came down to questions of value and valuation. Ah, but from where do we derive our values and valuations? I greatly valued this wildland and the experience it created for me. Leopold's land ethic, as a road map for escaping a two-dimensional view of the resource-conservation ethic, bubbled up through my consciousness that night. John stayed on the ridge to radio-track our cougar the next day. In the still, early morning light, I walked the ridge trail to the Rush Point fire tower on my way back to our base at Taylor Ranch on lower Big Creek. Spreading before and all around me for as far as I could see were the canyons and ridges of the Salmon River Mountains. Before me was a landscape that existed because conservation had been pursued within a conceptual framework of sustainable landscapes that included areas free of resource extraction. The prevailing resource-conservation paradigm would have blanketed the entire forest estate in sustained use. We need our wildland as well as our usable land.

Leopold implored us to value wilderness as a standard against which to judge the changes we make and their impact on landscapes. He didn't see wilderness as a flat mental construct. By contrasting material resources with ecological processes, he showed us how to escape the flatland of conservation ideas imposed on us by the progressive conservation movement. There were many contrasting, even polar, ecological constructs forming the basic architecture of his writing: birth-death, origin-evolution, equilibrium-non-equilibrium, predator-prey, ecological integrity–environmental decay and loss. In building an overarching land ethic, was Leopold not suggesting that we see landscapes in a new way, with one pole of our human-land vision anchored in the "wildland" and the other in the "urban" environment? In his vision, he was asking us not just to understand the wild and wilderness but also to understand and work with the middle grounds, those inherently unstable "middle landscapes" (Tuan 1998) in a more inclusive way. How can we understand and maintain the value of these middle landscapes unless we can appreciate and see them as more than a wellspring of resources and products? We must recognize that such landscapes form a complex and integrated system of interdependent processes and components, many under extreme stress.

In the foreword to *A Sand County Almanac*, Leopold wrote, "On this sand farm in Wisconsin, first worn out and then abandoned by our bigger-and-better society, we try to rebuild, with shovel and axe, what we are losing elsewhere. It is here we seek—and still find—our meat from God" (1949, viii). In an earlier, unpublished version of the foreword, Leopold lamented: "During my lifetime, more land has been destroyed or damaged than ever before in recorded history. As a field-worker in conservation, I have seen, studied, and measured many samples of this process. . . . During my lifetime, the thing called conservation has grown from a nameless idea into a mighty national movement. As a sportsman and naturalist, I have helped it grow—in size—but so far it seemed almost to shrink in potency" (1947). Here was hope and then despair, yet he breathed life into his human-land vision and left it to us to work out the details for ourselves in the places we work, and live, and love.

Living with Carnivores

Daily, I was gaining a sense of how the patterns and processes in this wildland shaped the lives of the cougars I was following (Seidensticker et al. 1973). I reasoned that the most important lesson I could gain from my wilderness experiences was to come to grips with the concept of wild. This seemed a central and defining idea in Leopold's human-land vision. Wild was a suture, an idea that linked most of what I was doing, thinking, and trying to understand at the time. I asked, By wild do we mean that this large, natural area was Nature preserved, untouched? Though I was living and working in the largest wilderness area in the contiguous forty-eight states, I often wondered what it would be like, what would be different about what I was seeing and measuring, if the assemblage of large carnivores was complete. Coyotes were contending with cougars, but what if grizzlies and wolves were extant? Or if the great anadromous fish runs of the past were still intact? Or if Native Americans still wintered on lower Big Creek?

This landscape in which I was young and working was a place of such great rugged beauty and wildness that I easily could overlook the meaning of the loss of its important pieces. But should I? Even here in the mountains dissected by the River of No Return, the scars and missing pieces pointed to humans as important past agents of change. Each time I flew for hours over this rugged landscape, searching for my radio-tagged cougars, I was aware that I was seeing only a snapshot of the landscape defined and restricted in time and space. The only way I could follow my cougars at this landscape scale was with radiotelemetry joined with a Cessna 185. For all its wildness, here was a land shaped by the interactions in the natural environment and, just as important, by its history of human use, facilitated by technological advancement and socioeconomic development.

The idea that our wilds, our wilderness areas, are strongly shaped by their history of human use is not a notion that resonates in the American mind (Nash 1973). Like the Romantic landscape painters (Raban 2001), we like our wilderness devoid of people. People get in the way of our notions of wild as pristine. But such notions can and do confound our understanding of the ecological patterns and processes that shape our wilderness benchmarks. I spent my second winter, a four-month block of my life, in this backcountry seeing and talking with only seven people. Some I liked; some I didn't. But I needed them. These people were an essential part of this wildland for me so I always had trouble with the idea that wild only had meaning without people.

We will likely never know how Native Americans changed the River of No Return landscape. A reading of a map reveals that there are few links or knowledge remaining from this distant past. Consider the names: Big Creek, Middle Fork, North Fork, South Fork, Johnson Creek. These are major rivers, and not a Native American name among them. Search the map and you won't find any. The wolves, and even the grizzlies, were extirpated before early miners and ranchers had time to note their presence (Hornocker 1970). By the turn of the century, deer and elk, bighorn sheep and mountain goat were nearly extirpated by market hunters. Mining created gashes in the headwater areas; great quantities of mercury were released into the streams as part of mining operations; the lower canyons were beaten down by the livestock that supported the mining industry. And, as Leopold (1949, 154) wrote, "cheat takes over" on the exposed hillsides. Introduced cheat grass was ubiquitous on the dry canyon slopes and ledges.

My only link to this historical past was through Wilbur Wiles and Jess and Dorothy Taylor, my wilderness mentors. They came into the country after the Second World War. Wilbur found the people, the remoteness, and the challenge of living here to his liking. He came looking for adventure and stayed on to trap and mine, though he never took more opal from his claim at a time than he could pack on two horses and in his backpack. He was fascinated by cougars and often tried to think like a cougar might think. He lived to watch and follow the interplay between his hounds and the tawny cats.

Jess, a ship's carpenter, found his mooring in this wild country. He and his young bride, Dorothy, took leave of the conventional life urged on them by family and friends and invested their life savings to purchase and live on the Lewis Place on lower Big Creek. With this small ranch as a base, they developed an exclusive sport-hunter clientele. Jess and Dorothy bet their ranch on Jess being able to find bighorn rams at the right season. He was successful.

I listened willingly to their stories about the men and women who had shaped the country before they arrived. A few old-timers were then still residing in out-of-the-way cabins and ranches. What emerged from their life stories and the stories they told me about the people they had known in their early years was a portrait of a changing ideology, changing human values that defined and shaped this wilderness over time just as much as humans had changed and shaped the natural patterns and processes in this wildland. Just as those who had gone before would not have guessed that Jess and Dorothy could make a living from high-end sport hunting or that Wilbur would make his living helping cougar researchers, I would never have guessed then that wolves would be reestablished here through a translocation effort thirty years hence.

Can we make a place for grizzlies here as well (Derr 2000)? Or will we continue to think and act as though grizzlies are out of place here? Is the River of No Return Wilderness too pastoral in our collective mindset to include grizzlies (Brownlow 2000)? Our changing values define our great wilds, our wildernesses. I like to believe that these wilds include grizzlies.

Landscape as History

Throughout *A Sand County Almanac*, Leopold lamented landscape loss and change, but he was at his best as an environmental historian. I don't know that he proposed criteria for the naturalness or the wildness of ecosystems such as degree of change, degree of sustained control, spatial extent of change, and abruptness of change (Angermeier 2000). But Leopold did tell us that "The image commonly employed in conservation education is 'the balance of nature' . . . this figure of speech fails to describe accurately what little we know about the land mechanism" (1949, 214). As an ecologist making observations on the scale of large landscapes, in his mind and writing Leopold replaced the equilibrium ecological perspective with a dynamic, nonequilibrium perspective. Further, I believe he would not have been surprised at the changing economic substrate that makes it possible for wolves and cougars to live in this landscape today, after having been so reduced in number and distribution a century ago. The West is shifting from dependency on an extractive economy to the Next West, a region not so dependent on resource extraction (Baden and Snow 1997). There is a lesson here for all of our efforts to conserve large carnivores in human-dominated landscapes everywhere.

In late summer, while tracking my cats in these headwaters of the Columbia River, I waded streams with only two or three spawning salmon. Wilbur remembered that twenty years before, salmon were so plentiful on their redds that you could almost walk across streams on their backs. I doubt that Leopold would have been surprised to learn the role anadromous fish have in moving nutrients from the sea to the nutrient-poor Columbia River system headwaters where I studied my cougars. Only a few percent of the marine-derived nitrogen and phosphorous (once delivered in the form of spent, dying salmon at the end of their annual runs) now reaches these streams (Barnard 2000) because salmon numbers are so diminished. I believe Leopold would have smiled at the connection, the linkage: salmon from the ocean, a keystone species in the wildlife and plant communities in these nutrient-poor headwaters.

I doubt, too, that Leopold would have been surprised that global climate change is threatening recovering populations of grizzlies wherever they occur in the intermountain West. Climate change is creating conditions where a fungus, blister rust, is increasing and killing whitebark pine trees (Matthews 2000). Whitebark pine nuts are a key food in the fall fattening cycle of the grizzly. Without this resource, grizzlies forage farther afield and, consequently, suffer higher mortality rates. This great carnivore is at risk from our actions far away from its immediate environs. Leopold's overarching land ethic vision recognizes that our wild areas are subject to cultural and ecological forces far beyond themselves; this makes drawing boundaries to include the ecological and cultural processes that sustain wild areas particularly challenging.

As an observant environmental historian, Leopold's vision of a land ethic was not so much about preservation as it was about our choosing how to direct inevitable changes in landscapes. My early confusion over Leopold's human-land vision came from a misconception. The vision I thought Leopold had imparted to me that summer I struggled with German was that of preserving pristine wilderness landscapes, free of human influence, where we come to see and learn from "the raw material out of which man has hammered the artifact called civilization" (Leopold 1949, 188). Later, I saw that Leopold's narratives are really about human history and how it has shaped the natural environment. Leopold realized that nearly every landscape is a cultural landscape, a landscape transformed. Leopold's vision fully recognized the importance of people and our options: our landscapes will be transformed, but if we understand and act in wise ways, we can influence how our landscapes will be changed.

Touching Wildness Every Day

I left the wilds of Idaho to learn about tigers in the Nepal *terai*, a ribbon of giant grassland and *Sal* forests at the foot of the Himalayas. By blending ancient and modern technology—chemical immobilization and radio-tracking conducted from the backs of trained elephants—I began to fill in some important gaps in our understanding of the ecology and behavior of this splendid great predator, the largest of the cats. Four times the mass of a cougar, a tiger is awesome. What I also found breathtaking was the large mammal assemblage living with tigers: rhinos and elephants, sloth bears and leopards, four species of deer, and wild pigs, all packed together at high density. This was as close as I was going to come to experiencing the drama of a large mammal assemblage such as those that existed during the Pleistocene. This was the Pleistocene assemblage walking under my tree-blind each day and night. This was wild. But as wild and remote as my study site in the Royal Chitwan National Park was in those days, I was never far from people. I could nearly always listen to the gentle sounds, the tempo of village life just across the river from my tree-blinds. I thought of the American concept of wilderness being devoid of people. How foreign such an idea of wild seemed here among tigers, and rhinos, and *Tharu* people. Surrounding my tree-blinds were old farm sites. Yet, deer came here to graze; tigers came here to kill deer. Leopold's overarching human-land vision seemed entirely appropriate to me in the Chitwan Valley. In subsequent years, it has been working for the benefit of people and wildlife (Seidensticker et al. 1999).

Time passed. I found myself working at the Smithsonian's National Zoological Park in Washington, D.C. Although the establishment and maintenance of great wilderness areas remained important to me, my interests shifted to possibilities closer to home. Ian McHarg (1969) noted that the physiographic expression of Washington, D.C. begins on the flats where a great city meets a great river. The forested corridors extend out from the flats "like the ribs of a fan," linking coastal plain with the Piedmont and beyond. I looked for and discovered wildness in this big city. What became important for me was that I could walk or bike to work. When I reached the bend in Rock Creek in the zoo near the spectacled bear exhibit, I had two choices: I could turn up Beaver Valley past the Mexican wolf habitat (where we will breed and prepare Mexican wolves for eventual reintroduction into the Aldo Leopold Gila Wilderness Area in New Mexico) on to my office to work on tiger conservation in Asian landscapes, or I could turn down Rock Creek, take the C & O Canal Towpath to Harper's Ferry, turn south on the Appalachian Trail, and walk to the Smoky Mountains and see the thin green slice that remains of the great eastern deciduous forest.

The zoo is part of a great network of greenways that provides the opportunity to become immersed in a great wild experience—or a local adventure scaled to time available. Not as wild as the River of No Return, but wild nonetheless. I was coming to understand wildness in the ways Leopold meant us to, I believe. Creating greenways underlies a process of creating an outdoor ethic, a personal involvement in the outdoors as an essential part of life in the context of the culture and economy of our times. I live in a landscape transformed. If we are learning from our environmental history, we have options on the extent and nature of those changes. Our land ethic—wildness in our minds—will direct these changes.

This is getting closer to Leopold's vision, I think. This is why he enjoyed the sand counties and his farm. Humans had been heavy-handed in the sand counties, but Leopold saw options for recovery if people could choose to see themselves, not as distinct from land, but as visionaries and agents of change in recovering landscapes; here is wildness close to home, a wildness we touch every day.

Susan, my wife, and I became increasingly infatuated with our garden, a forty-by-forty-foot former backyard lawn in urban Washington. When we moved into our ninety-year-old house a dozen years ago, Susan declared that she wanted a wild, romantic, American garden, not grass and a garage, as her backyard. The scale of my interests in wildness was contracting, but my understanding of wildness was expanding. We hauled rocks, dug ponds, and planted wildflowers. We counted birds and plants and compared these numbers with those of other similarly sized patches of wildness. As wildness took over, our garden took on a stronger and deeper romantic atmosphere. Upon the death of our old dog, squirrels, raccoons, and Virginia opossums increasingly found refuge there. The loss of our domestic predator tipped the direction of change in the inherently unstable middle landscape of our garden. There, at a scale, we could feel and smell and see; we had direct, personal experience of the processes that shaped wildness, the way nature and our minds work. First for Thoreau, and then for Leopold, wildness was a state of mind rather than a description of a place. Both men championed a land *and* people ethic and not a land versus people ethic. Humans are a central environmental force shaping landscapes everywhere. Wildness is our touchstone, "a question of intellectual humility . . . (giving) definition and meaning to the human enterprise" (Leopold 1949, 200, 201).

Acknowledgment: Wildness does not necessarily imply aloneness. Thank you, Susan, for exploring wildness with me.

References

Angermeier, P. L. 2000. The natural imperative for biological conservation. *Biological Conservation* 14:373–381.

Baden, J. A., and D. Snow, eds. 1997. *The next west: public lands, community, and economy in the American West.* Washington, D.C.: Island Press.

Barnard, J. 2000. Salmon decline hurts many other species. *Seattle Daily Journal of Commerce*, 8 February. www.djc.com/news/enviro/11003918.htm

Brownlow, A. 2000. A wolf in the garden: ideology and change in the Adirondack landscape. In *Animal spaces, beastly places: new geographies of human-animal relations*, ed. C. Philo and C. Wilbert, 141–158. London: Routledge.

Derr, M., 2000. Grizzly bears poised to make comeback. *The New York Times*, May 30, D:1, 4.

Foster, D. R. 1999. *Thoreau's country: journey through a transformed landscape.* Cambridge, Mass.: Harvard University Press.

Hays, S. P. 1959. *Conservation and the gospel of efficiency: the progressive conservation movement 1890–1920.* Cambridge, Mass.: Harvard University Press.

Hornocker, M. G. 1970. An analysis of mountain lion predation upon mule deer and elk in the Idaho Primitive Area. *Wildlife Monographs* 21:1–39.

Leopold, A. 1947. Unpublished foreword to *A sand county almanac and sketches here and there.* In *War against the wolf: American campaign to exterminate the wolf*, ed. R. McIntyre, 321–325, 1995. Stillwater, Minn.: Voyageur Press.

Leopold, A. 1949. *A sand county almanac and sketches here and there.* New York: Oxford University Press.

Matthews, M. 2000. Western pine species loses ground to neglect. *The Washington Post*, 27 March, A:11.

McHarg, I. L. 1969. *Design with nature.* Garden City, N.Y.: Doubleday.

Nash, R. 1973. *Wilderness and the American mind.* 2d ed. New Haven, Conn.: Yale University Press.

Raban, J. 2001. Battle ground of the eye. *The Atlantic* 287(3):40–52.

Seidensticker, J., M. G. Hornocker, W. V. Wiles, and J. P. Messick. 1973. Mountain lion social organization in the Idaho Primitive Area. *Wildlife Monographs* 35:1–60.

Seidensticker, J., S. Christie, and P. Jackson, eds. 1999. *Riding the tiger: tiger conservation in human-dominated landscapes.* Cambridge: Cambridge University Press.

Thoreau, H. D. 1947. *Walden.* In *The portable Thoreau*, ed. C. Bode, 258–572. New York: Viking Press.

Tuan, Yi-Fu. 1998. *Escapism.* Baltimore, Md.: Johns Hopkins University Press.

Chapter 4

Then and Now

JAMES A. ESTES

To be entirely honest, Aldo Leopold was unknown to me in 1967 when I began graduate study in wildlife biology at Washington State University. I entered graduate school with a traditional education in zoology, which did not emphasize conservation, natural resource management, nor, for that matter, even ecology. Frankly, I can't remember what it was that drew me to the natural sciences other than perhaps a fondness for nature and an aversion to normal day-to-day work activities. I first learned of Aldo Leopold from Professor Irvin Buss, one of Leopold's last doctoral students at the University of Wisconsin. *A Sand County Almanac* was recommended reading in Professor Buss's wildlife ecology course, which I took during my first semester. The course itself probably left little of substance in my worldview, but my first reading of *A Sand County Almanac* struck an emotional chord that I will carry to my grave. From the initial sentence of the foreword, I counted myself among those who could not live without things that are wild and free. Although biology was a topic of boundless wonder for me in those days, I knew then that my quest was for a life that offered intellectual fulfillment in places that were wild and free. In looking back now, the poetic force of Leopold's words influenced both the way I think about nature and my approach to science. This essay is a reflection on those influences.

Leopold's Visions of Nature

A Sand County Almanac has touched the human soul in various ways. For some, the power of Leopold's writing lies in its poetic quality—his magical ability to capture in words the emotions that most of us only feel when we hear a wolf howl or see the spectacle of sun and clouds at twilight. For others, this power comes from the philosophical revelation of his land ethic—a neo-Malthusian realization that humans must learn to value land as the cradle of existence rather than as the economic commodity of further growth and development. For a smaller number of professional ecologists, Aldo Leopold's force lies in his remarkable grasp of nature's fabric. I have read *A Sand County Almanac* on various occasions over the past thirty-four years, usually in an effort to revitalize my own spirit. But in refreshing my memory of the details as I prepared to write this essay, I was astonished to discover that Leopold had observed and clearly understood the ecological relationships among wolves, ungulates, and vegetation more than fifty years ago—that the systematic elimination of wolves from North America meant too many ungulates and overgrazed rangelands. I really can't imagine how this important realization escaped me, as it apparently has escaped so many others who also should know better.

The roots of this general idea—that herbivores are limited by predators, thus releasing plants from regulation by herbivory—are commonly attributed to Hairston, Smith, and Slobodkin's Green World Hypothesis, published in 1960 and still a cornerstone in the conceptual arsenal of community ecology (Hairston et al. 1960). Food chain interactions of this sort have come to be known as trophic cascades (Carpenter and Kitchell 1993), and the degree to which populations are regulated by "top-down control" remains one of ecology's foremost topics of debate and inquiry, especially in terrestrial ecosystems (Pace et al. 1999; Polis et al. 2000). In the early 1990s, newly discovered evidence for a trophic cascade among wolves, moose, and balsam fir on Isle Royale in Lake Superior was viewed as so novel and important that it warranted publication in *Science* (McLaren and Peterson 1994).

But having lived through that time in history when wolves went from being abundant to rare, Aldo Leopold saw and understood the general process much earlier.

The conceptual significance of Leopold's allegory on wolves and deer escaped me in 1968, even though eight years already had passed since publication of the Green World Hypothesis, during which time the subject of top-down control in population regulation was an issue of active interest and debate (Murdoch 1966; Ehrlich and Birch 1967). My failure to make the connection probably stemmed from the fact that I had not yet seen enough of nature firsthand to solidify the ideas in what was then a contextually vacant mind. However, the fact that the centerpiece of my life's work on sea otters and kelp forests was to be founded on this very concept strikes me in retrospect as ironic, if not somewhat embarrassing.

Although I admit to being unmoved by the Green World Hypothesis during those early years and blind to the significance of Leopold's vision, a fascination with the larger temporal dimensions of nature was beginning to capture my lasting interest. Every person with even a modicum of sense has a clear grasp of what time is. But in the *Almanac*'s entry for February, the poetic chronicle of time that unfolded as Leopold's sawyer passed through the "good oak's" eighty annual rings made the interplay between time and nature come alive for me. Considerations of history and time were not new to biology. Time had long been a fundamental dimension in paleontology and evolutionary biology, but wildlife biologists and ecologists tended to collapse the relevant temporal dimensions in their thinking to scales that could be observed over the course of a human lifetime, usually much less time. This limited perspective of time probably stemmed in large part from a desire for methodological rigor in ecology and wildlife biology. Many practitioners of these disciplines believed then that the best science comes from controlled experiments, which necessarily limit the scale of time over which science can be done. But this approach also leaves many important questions unanswered and unanswerable. It also tends to lull us into a state of intellectual apathy when confronted with the potential importance of past events, such as the extinction of wolves in large parts of North America. Moreover, such an approach encourages us to view modern-day ecosystems as though they have always been more or less the way we now understand them to be, even though we all know that isn't the case. I admit that this characterization of how many ecologists viewed the world is an oversimplification and perhaps even unfair, but one need not look far for examples of its overlying truth.

Why History Matters

History matters for the simple reason that it is the pathway to the present. This principle is well known in both human societies and the field of evolutionary biology, where fortuitous events have set the course for all that follows. Imagine just how different the track of human history might have been if the Mesopotamians had not lived in a region with such a large proportion of the world's potentially domesticable plants and animals. The resulting shift from a hunter-gatherer to an agricultural society permitted the accrual of wealth (largely in the form of food), in turn leading to the development of armies and European dominance in the course of world conquest (Diamond 1997). Imagine just how different life on earth would be today if a happenstance meteor had not collided with Earth some sixty-five million years ago, extinguishing most of the dinosaurs and setting the table for the Cenozoic radiation of birds and mammals. In neither case is it possible to predict the course of events had the original natural phenomenon not occurred. But without doubt, subsequent history would have been very different.

The fundamental importance of many other signal events in human history and evolutionary biology is widely known and broadly appreciated. However, the conceptual underpinnings of ecology have never placed an unequivocal premium on the nature of connections between past and present. Lewontin (1969) set the stage for thinking about this issue in his classic paper, "The Meaning of Stability," in which he recognized two fundamentally different ways that ecological systems might behave. Lewontin imagined that systems perturbed from equilibrium conditions are subject to "transition vectors"—forces that cause populations to change depending on their altered relationships with other species and the physical environment. For instance, an abundant population of deer reduced by overhunting or a hard winter would be expected to increase until it returned to its natural equilibrium state; the transition vector in this case would be the demographic differential between births and deaths, fueled by a surplus of food. But Lewontin also recognized that whether perturbed systems return to their initial state or some other state depends on whether the field of transition vectors is linear or nonlinear. If the transition vectors are linear, the system will always return to its initial state. However, if the transition vectors are nonlinear, a perturbed system may shift to alternate stable states, depending on the magnitude and direction of the perturbation. History is of no real consequence to globally stable systems because such systems predictably return to the same state following perturbations of all kinds, so long as no species is extinguished in the process. However, history is an essential dimension in understanding systems with multiple stable states because their current position always depends on where they came from. Although there now is reasonably good evidence for the existence of systems with multiple states in nature, ecologists remain undecided as to whether such behavior is the exception or the rule. Thus, even from this theoretical perspective, the importance of time and history in understanding ecosystem behavior is equivocal in the minds of many ecologists.

Some Examples

A growing number of examples point to the depth of understanding that historical data bring to wildlife biology, fisheries, and ecology. The foremost such example in my mind is in what Janzen and Martin (1982) have termed "neotropical anachronisms"—alleged misinterpretations by many tropical ecologists of the results of modern-day studies by failing to recognize important roles played by the recently extinct New World megafauna. Janzen and Martin's insightful theory was founded on a history of observations of these facts: numerous tropical trees produce large, sweet fruits that surround tough seeds, and, throughout much of the modern neotropics, these fruits fall to the ground below their parent trees where they rot or are eaten by various micro- and mesoherbivores. The dynamics of these fruit-herbivore interactions had been the subject of a rich ecological literature in which, in many cases, evolutionary consequences were stated or inferred. Janzen and Martin speculated that, in the past, these fruits were likely consumed and their seeds dispersed by the gomphotheres and other large Pleistocene frugivores. Thus, they rarely, if ever, accumulated to rot or be consumed by the micro- and mesoherbivores until the New World megafauna was extinguished some ten to fifteen thousand years ago. They used changes in plant distribution and fruit-frugivore dynamics associated with the recent reintroduction of Pleistocene horses and other large mammalian herbivores to the New World tropics to make their point.

Similar stories are unfolding through the amalgamation of ecological and retrospective data to enrich our understanding of other contemporary ecosystems. For instance, Jackson et al. (2001) have concluded that coastal marine systems throughout the world now function in grossly different ways than they did in the recent past, in large part owning to the reduction or extinction of large consumers and other abundant species. For example, in the Chesapeake Bay, the removal of oyster reefs (estimated to have filtered a volume of seawater equivalent to the entire bay every three days!) has brought about dramatic changes in the marine food web, converting a system driven by benthic producers into one driven by phytoplankton. Benthic cores provide marine ecologists with evidence that the waters of the Chesapeake were clear before the oyster reefs were gone. Similarly, notes recorded by early Spanish explorers reveal that Caribbean sea grasses were short and sparse compared with what they are today. This difference is likely connected to the history of sea turtles and perhaps manatees, which consume sea grasses. These animals were spectacularly abundant several centuries ago but are rare today. In these and many other cases, the knowledge of history adds richness and relevance to the results of modern-day research in ecology, wildlife, and fisheries. These connections, for the most part, are yet to be discovered.

The Tree-Ring Prophecy

The aforementioned examples, although probably conceived by their creators with little or no influence from Aldo Leopold's words in *A Sand County Almanac*, nonetheless are indicative of the prophetic nature of his thinking. Nowhere is that prophetic insight more evident than in his prediction that "Some day some patient botanist will draw a frequency curve of oak birth-years, and show that the curve humps every ten years, each hump originating from a low in the ten-year rabbit cycle" (Leopold 1987, 7). Indeed, tree rings have since been widely used by ecologists and geochronologists to unlock the broader sweep of nature's time, including the history of fire and rainfall and even the influence of predators. McLaren and Peterson (1994) used tree-ring analysis of the growth of balsam fir saplings on Isle Royale to chronicle evidence in support of their theory of top-down control by wolf predation on moose. Similarly, Ripple and Larsen (2000) reconstructed a frequency curve of aspen birth years from the record of annual rings in living trees to show that there has been virtually no recruitment in northern Yellowstone aspen forests since about 1920, the approximate year that wolves were extinguished from Wyoming to better the interests of ranchers and big-game hunters. To use Leopold's own words, these examples "attest the unity of the hodge-podge called history" (Leopold 1987, 16).

How many similar stories will emerge as future testimonies to Leopold's vision of time? Many, I suspect. Consider the potential significance of history to the understanding of a major human epidemic, Lyme disease. The ecology of Lyme disease is well known, at least in the northeastern United States (Jones et al. 1998). The disease is caused by a microbe contracted by humans through the bite of deer ticks. Deer ticks require two other species —white-footed mice and white-tailed deer—to complete their life cycle. If either of these species is rare, tick populations remain depressed and the threat of Lyme disease in humans is reduced. Oak mast events are the key to bringing sufficient numbers of deer and mice together in space and time because both species are attracted to acorns. Consequently, the threat of Lyme disease is greatest following oak mast events, which occur episodically in eastern deciduous forests. Now imagine how the events of history may have impinged on this situation. Oaks and chestnuts are natural competitors in forest ecosystems. Oak forests undoubtedly have become more abundant over the past century because of chestnut blight. White-tailed deer also have become more abundant in recent decades with the extinction of wolves, their main predator. Today, mast crops probably accumulate in larger numbers and remain on the ground longer than they did when passenger pigeons darkened the skies of the eastern United States. In fact, this now extinct species appears to have been designed to key in on and exploit oak mast crops, traveling from area to area in immense numbers in search of these events. Might it be that the collective force of these events has lead to the Lyme disease epidemic?

Nature's Fabric

The fabric of nature is woven more by species interactions than by species themselves. But while species are reasonably easy to observe, species interactions are not. Witness the large number of species and the small number of species interactions that are presently known to science. The cryptic nature of species interactions stems from the fact that such interactions are typically only recognized following perturbations to one of the interactors. This is one of the principal reasons that ecologists do experiments—to clarify the nature and strength of species interactions. The experimental approach has indeed revolutionized ecology. The problem is that the amount of time needed to see an interaction varies greatly among species and ecosystems, in general scaling to generation time, thus explaining why so much of experimental ecology has focused on short-lived species and why so little is known about interactions among longer-lived species. Many species studied by traditional wildlife biologists are experimentally intractable.

A good deal of my professional life has been devoted to the study of interactions between sea otters and kelp forests, an experimentally intractable system but one in which history and time have served as essential methodological allies. I was introduced to sea otters and kelp forests at Amchitka Island in 1970, a time at which sea otters were abundant and had been for several decades following their recovery from near extinction by overhunting in the Pacific maritime fur trade. I remember studying the otters and diving in their kelp forest habitat for nearly a year with a mind toward somehow understanding the ecology of that system but with almost no idea of what to look for. I was aware that, in the past, otters had been eliminated from nearly all of the Aleutian Islands and were still absent from some of these, but my mind was so focused on bottom-up forcing that it never occurred to me that a visit to one of the unoccupied islands might be of any interest. In 1971, I was fortunate to meet R. T. Paine, who had recently completed the first of his now classic studies of predation by sea stars in intertidal mussel beds. Bob encouraged me to think about how the otters might be influencing the kelp forests. With that kernel of inspiration, a reason to visit the unoccupied islands was clear and urgent.

So, together with fellow graduate student John Palmisano, I traveled to Shemya Island (then unoccupied by otters) later that summer for a look. Our first peak at the sea floor immediately provided one of the most exciting revelations of my professional life. While Amchitka supported a few small sea urchins and lush kelp forests, at Shemya, large and abundant sea urchins had grazed away the kelp forests. Knowing that sea otters consumed sea urchins, that sea urchins consumed kelp, and that sea otters had once been abundant at Shemya, we immediately surmised that sea otters played an essential role in maintaining the kelp forests (Estes and Palmisano 1974). Further investigations comparing areas where sea otters had recovered with those from which they remained absent confirmed this interaction, as have longitudinal studies of particular sites at which the otter numbers subsequently have increased or decreased. This same approach has revealed complex linkages between sea otters and numerous other species in the kelp forest ecosystem, thus contributing to our view of the sea otter as a keystone species. This experience taught me the role history could play in shedding light on natural processes that might otherwise be difficult or impossible to discern.

Concluding Remarks

My view of nature has been strongly influenced by the force of history, in part from what I have seen in my own work and in part from what I have learned from others. It is difficult for one to look back over a lifetime and really understand how such views come to be, although there are formative events that stand out. Aldo Leopold's chronicle of time as the sawyer's blade passed through eighty years of history in the "good oak" was one such event in my life. Having read *A Sand County Almanac* several times over the course of my professional life, most recently in preparation to write this essay, I am struck more deeply at each reading by how much Leopold's view of nature was guided by an interest in and understanding of the broader sweep of time. Aldo Leopold is widely acknowledged as a visionary in the field of conservation. His appreciation for the intimate connections between time and nature is no less prophetic.

References

Carpenter, S. R., and J. F. Kitchell, eds. 1993. *The trophic cascade in lakes.* New York: Cambridge University Press.

Diamond, J. 1997. *Guns, germs, and steel.* New York: Norton.

Ehrlich, P. R., and L. C. Birch. 1967. The "balance of nature" and "population control." *American Naturalist* 101:97–107.

Estes, J. A., and J. F. Palmisano. 1974. Sea otters: their role in structuring nearshore communities. *Science* 185:1058–1060.

Hairston, N. G., F. E. Smith, and L. B. Slobodkin. 1960. Community structure, population control, and competition. *American Naturalist* 94:421–425.

Jackson, J. B. C., M. X. Kirby, W. H. Berger, K. A. Bjorndal, L. W. Botsford, B. J. Bourque, R. Bradbury, R. Cooke, J. A. Estes, T. P. Hughes, S. Kidwell, C. B. Lange, H. S. Lenihan, J. M. Pandolfi, C. H. Peterson, R. S. Steneck, M. J. Tegner, and R. Warner. 2001. Historical overfishing and the recent collapse of coastal ecosystems. *Science* 293:629–637.

Janzen, D. H., and P. S. Martin. 1982. Neotropical anachronisms: the fruits the gomphotheres ate. *Science* 215:19–27.

Jones, C. G., R. S. Ostfeld, M. P. Richard, E. M. Schauber, and J. O. Wolff. 1998. Chain reactions linking acorns to gypsy moth outbreaks and Lyme disease risk. *Science* 281:347–351.

Leopold, A. 1987. Special commemorative edition of *A sand county almanac and sketches here and there.* New York: Oxford University Press.

Lewontin, R. C. 1969. The meaning of stability. *Brookhaven Symposia in Biology* 22:13–24.

McLaren, B. E., and R. O. Peterson. 1994. Wolves, moose, and tree rings on Isle Royale. *Science* 266:1555–1558.

Murdoch, W. S. 1966. Community structure, population control, and competition—a critique. *American Naturalist* 100:219–226.

Pace, M. L., J. J. Cole, S. R. Carpenter, and J. F. Kitchell. 1999. Trophic cascades revealed in diverse ecosystems. *Trends in Ecology and Evolution* 14:483–488.

Polis, G. A., A. L. W. Sears, G. R. Huxel, D. R. Strong, and J. Maron. 2000. When is a trophic cascade a trophic cascade? *Trends in Ecology and Evolution* 15:473–475.

Ripple, W. J., and E. J. Larsen. 2000. Historic aspen recruitment, elk, and wolves in northern Yellowstone National Park, USA. *Biological Conservation* 95:361–370.

Chapter 5

Great Possessions

Leopold's Good Oak

MARY ANNE BISHOP

Hindsight was among Aldo Leopold's greatest literary attributes. Throughout *A Sand County Almanac*, he reflects on his personal outdoor experiences and their application to wildlife management and conservation. Nowhere, however, are the historic merits and mistakes of wildlife management more clearly chronicled than in his essay "February Good Oak." Leopold's biographer, Curt Meine (pers. commun.), considers "Good Oak" "[a] litany . . . a roll call of significant indicators of landscape change, implicitly making the point that the land and its components do indeed change, and history's job is to bring some unity and continuity to the record."

In "Good Oak," Leopold and his wife, Estella, the "chief sawyer," saw through eighty annual rings of an oak whose end was delivered by a stroke of summer lightning. Leopold uses this "integrated transect of a century" to provide a concise picture of Wisconsin's conservation history against the backdrop of nature's powerful forces. Beginning with his recent acquisition of the farm, Leopold describes the moods and events of each passing year and decade. He ends at the oak's pith year, 1865, the year John Muir offered to buy from his brother his boyhood farm for "a sanctuary for the wildflowers that had gladdened his youth," a year that "still stands in Wisconsin history as the birth year of mercy for things natural, wild and free" (Leopold 1949, 15–16).

Readers may want to return to Leopold's "Good Oak" before reading this essay. Although Leopold had a keen sense of wildlife's historic fate, and expressed it with unique eloquence, I sometimes wonder if even he could have foreseen all that has happened to "wild things" over the last half of the twentieth century. Did Leopold assume that Malthus's prophesied pace of human population growth would continue? Did Leopold foresee our country's imminent economic boom and all of its environmental costs, including an onslaught of synthetic chemicals and widespread habitat losses that would alter wildlife populations no matter what wildlife policies were adopted? Did Leopold foresee the rise of a popular environmental movement and a politically influential backlash by those who would co-opt the term "wise use"?

How would Leopold have chronicled these past fifty years? What follows is one person's interpretation of the course of wildlife management since *A Sand County Almanac* was published. This essay highlights efforts, both fruitful and barren, to manage and conserve "our fellow voyagers . . . in the odyssey of evolution" (Leopold 1949, 109). Can we learn from the lessons of the "Good Oak" and apply Leopold's enduring wisdom as we rush into the new millennium?

The Leopold Oak

Here is the story of the Leopold oak, a lucky bur oak acorn that germinated amidst a prairie sea and soon after was transplanted by "The Shack" on the Leopold farm, near an oak planted earlier to honor the distinguished ecologist Charles Elton. Planted to honor Leopold, this good oak will never warm February's hearth, lightning and disease withstanding. But this oak's history is "something more than wood." It is the integrated transect of a half century of Leopold's great legacy to conservation and wildlife management, "written in the concentric annual rings of good oak" (Leopold 1949, 9).

As the acorn germinates in spring 1948, DDT is sprayed across the land. Farmer, gardener, and mosquito controller discover its efficiency and wildlife biologists track its devastating effect on birds, fish, and mammals. The seedling takes root as eighteen billion board feet of its fellow trees are sold, a fifty-year sale (Rakestraw 1981) that signals the onslaught of industrial harvesting of old-growth trees in the Tongass, our largest National Forest. In fall 1949, when the seedling is transplanted, it does not hear the first cannon net trapping waterfowl flocks, the unanswered gobbler calls of hunters in nineteen states (Lewis 1987), or the last calls of one hundred thousand mallards in Illinois that are dying from lead-shot ingestion (Ehrlich et al. 1988).

"Spray!" cry the farmers, and we pause for breath.

Rabbit negligence prevails and the two-year-old oak survives into the second half of the century. The country braces itself for fifty more years of rapid urbanization, habitat destruction, computerization, global warming, and the demise and comeback of wildlife populations. What is happening to your favorite haunt? Will it survive the bulldozer? Escape the plow? The country supports 152 million humans, while the world's whooping crane population is 31 (McNulty 1966) and Wisconsin's wolf population is 50 (Thiel 1993). Federal aid for fish becomes a reality under the Dingell-Johnson Federal Aid to State Fisheries Act. States are now funded to restore, manage, and introduce sport fish into newly acquired or developed lakes. The following year, the Federal Aid in Wildlife Restoration (Pittman-Robertson) Act is saved from the vagaries of annual appropriations when Congress enacts a permanent, indefinite appropriation status (Williamson 1987). The oak is still thirty-three years short of producing its first acorns for quail and squirrels when a comprehensive reference on wildlife food habits is published, summarizing seventy-five years of studies and declaring acorns "the staff of life for many wildlife species" (Martin et al. 1951, 308).

In 1953, the country takes one step forward and three steps backward as another fifty-year contract is awarded on our largest national forest. A year later, a forty-one-year-long harvest of over 400,000 acres of old-growth forest begins (USDA Forest Service 1996). But if the oak hears the ancient trees fall, its wood gives no sign. It grows on while the Department of the Interior grants 274 grazing, logging, mining, and oil drilling leases on refuge lands between 1953 and 1955 (Matthiessen 1959). It is four feet tall in 1954 when California's wetlands weigh in at a lean 560,000 acres, down from almost 6 million acres that once covered this western state (Horn and Glasgow 1964). That same year, the remaining breeding ground for whooping cranes is accidentally discovered in a remote section of northwest Canada.

In 1955, the oak's branches reach for the sky as a small plane flies overhead conducting the first North American waterfowl breeding population survey (Bellrose 1980) for the country's twenty-five million sportsmen (Belanger 1988). Also in 1955, a fledgling organization known as The Nature Conservancy purchases its first tract of land. In 1957, the oak stands six feet tall as subsidized farmers revert wheat fields back to their natural state, natural except for the sixty-five thousand tons of dieldrin, heptachlor, and other persistent chemicals descending from the air onto the fields and forests (Matthiessen 1959). Wisconsin's bounty for wolves is repealed, but the oak does not notice that it is already too late for Wisconsin's wolves. In 1959, the oak is one foot taller as Alaska bans fish traps in its salmon-laden waters (Scudder 1970) and hunters celebrate the return of the turkey gobbler (Lewis 1987). The oak's growth is slowed down by the year's severe drought, which forces thousands of prairie pothole ducks to forego nesting.

"Timber!" cry the lumberjacks, and we pause for breath.

As it enters the 1960s, the oak is water stressed by the continuing drought. Wildlife biologists are armed with The Wildlife Society's *Manual of Game Investigational Techniques* (Mosby 1960), dart guns, radio transmitters, and computer punch cards—tools that help them to uncover the lives of our most secretive, cryptic, and mobile animals. By 1961, prairie duck populations are ravaged by the drought, and the lowest daily bag limit on ducks ever, two per day, is mandatory in the four flyways (Hawkins 1980). Drought and oak withstanding, business prevails: nearly 33,500 tons of DDT are sprayed. Many of these same prairie ducks will consume and bioaccumulate the chemicals (Buckley and Springer 1964).

The oak grows in 1962, when Rachel Carson publishes *Silent Spring*. The book chronicles the effects of persistent pesticides such as DDT. "Our fate is connected with the animals" the author writes (Carson 1962). This same year, wildlife values in wetlands must be considered prior to drainage (Burwell et al. 1964). The oak tops eleven feet in 1964 as our "Great Society" president sets aside millions of acres of wilderness, the last of our greatest lands. But the oak does not notice that the peregrine falcon no longer flies over many of these same lands, its numbers in decline because of DDT.

The oak grows in 1965 when Colorado declares that the mountain lion is no longer vermin, but game. Other western states are prompted to follow suit (Hornocker and Quigley 1987). In 1966, an indigo bunting and an eastern kingbird sing from the good oak's branches during the first Breeding Bird Survey, while in Texas the once one-million-strong Attwater's prairie chicken is declared endangered (Stap 1996). The prairie chicken joins a medley of vertebrate species and subspecies teetering on the brink of extinction: twelve mammals, eighteen Hawaiian and fourteen non-Hawaiian birds, two reptiles, two amphibians, and fifteen fish (USDI Fish and Wildlife Service 1983). Does the oak heed the plight of the blue pike? The Texas blind salamander? The blunt-nosed leopard lizard? Kirtland's warbler? Does the oak notice that the prairie bush clover and eastern prairie fringed orchid no longer bloom nearby? Their fates are now in the hands of federal agencies that have been encouraged to consider their protection.

The oak grows in 1967, when the folly of a proposed 6.9-million-acre reservoir, on the site of Alaskan breeding grounds for 1.5 million ducks, is permanently shelved (Coates 1991). By 1969, the oak is eighteen feet tall. Antiwar protesters are marching in nearby Madison, and an innovative book called *Design with Nature* (McHarg 1969) is laying new groundwork for spatial analysis in resource management. The Endangered Species Act is getting stronger, and the landmark National Environmental Policy Act calls for full evaluation of environmental impacts and consequences for federally financed projects. The oak grows, but can America's biodiversity meet and endure us, the enemy?

"What's it worth?" cry the politicians, and we pause for breath.

In 1970, the oak is nineteen feet tall as twenty million Americans march in demonstrations, attend teach-ins, and participate in cleanups in celebration of the first Earth Day. Freshwater is once again flowing across the Florida Everglades' River of Grass, slowly restoring its fish nurseries and wading bird habitat (Tiner 1984). The oak grows in 1971, a record drought year in Florida, when the state drains 140,000 acres of wetlands to straighten the meandering Kissimmee River into the new Canal C-38 (Graham 1987). In 1972, the oak adds its usual ring as the federal government restricts use of DDT, a probable carcinogen with adverse impacts on fish and wildlife, and bans Compound 1080, the poison most responsible for killing prairie dogs and coyotes. In the same year, the first earth observation satellite is launched, heralding a new tool to analyze and monitor landscapes at a regional level. The Clean Water Act is passed, requiring permits to fill wetlands. Hurricane Agnes hits Chesapeake Bay in June, burying grass beds with runoff sediment from industry, agriculture, and cities, thus destroying the winter habitat of thousands of canvasbacks and redheads (Tiner 1984).

The oak grows in 1973, as the president of the United States reaffirms our belief that "nothing is more priceless than the rich array of animal life with which our country has been blessed" and signs a more effective Endangered Species Act ordering participation by all federal agencies. The oak grows in 1974, when the Fish and Wildlife Service establishes the National Wetlands Inventory Project and takes its first steps in the long struggle to replace lead waterfowl shot with nontoxic steel. Also in 1974, a group of Mexican conservationists recognizes their country's importance to wintering waterfowl and creates Ducks Unlimited de Mexico. In 1975, the oak does not hear the pack of wolves once more howling in northern Wisconsin (Thiel and Welch 1981). In the same year, Nebraska abandons aerial waterfowl breeding surveys in its few remaining Rainwater Basin wetlands.

The oak reaches twenty-three feet tall in 1976, a year of landmark legislation. Florida begins restoration of the Kissimmee River, and the United States bans manufacturing of PCBs. That same year, the Forest Service begins two new tasks: providing for the diversity of plants and animals and managing habitat for viable populations of native wildlife. But the oak does not notice the unspoken questions: how many gray bats, Alexander Archipelago wolves, Florida panthers, and Maui parrotbills are just enough? How much prairie is just enough for the mountain plover, Sprague's pipit, Swift fox, and black-tailed prairie dog? What size forest fragments and how many acres of old growth are just enough for the fisher, northern spotted owl, red-cockaded woodpecker, and marbled murrelet? How much stream flow is just enough for the humpback chub, Umpqua cutthroat trout, and northern California's coho salmon? How much wildness is just enough for the grizzly bear, ocelot, and gray wolf? And just how many species will suffice for our children's children to enjoy?

The oak grows in 1977, when Colorado introduces the first income tax checkoff to support nongame and endangered species (Anderson 1987) and our president calls upon federal agencies to minimize the loss and degradation of wetlands. In 1978, webworms defoliate parts of the oak's crown. Loons no longer nest on many New York lakes; the likely culprit—recreational boating (Ehrlich et al. 1988). That same year, a court awards $7.5 million in compensation to mitigate impacts of an upstream dam and charges a conservation group to protect and maintain habitat for the millions of migratory birds in Nebraska's Platte River (Lingle 1992). The oak's buds burst open two weeks early in 1979, when a future Forest Service chief publishes a book that details impacts of forest management on wildlife habitat and provides a system of classifying wildlife habitats according to both their land features and their natural vegetation (Thomas 1979).

"Save them!" cry our children, and we pause for breath.

Our oak lays down its ten rings during the 1980s, unaware of the decade's rapid technological changes and the overflowing paper-recycling bins. Each new computer innovation is rapidly replaced by a faster, improved operating system, software program, or other new technological wizardry. Wildlife managers enter the decade equipped with programs that, within minutes, allow them to determine survival rates, estimate population sizes and densities, project harvest rates, map home ranges, and model populations and habitat suitability. LANDSAT satellites are fast becoming tools of the trade, providing high-resolution imagery and the capability to monitor movements of grizzly bears and sandhill cranes. By middecade, GIS (Geographic Information Systems) joins the ranks of wildlife acronyms and is mandatory methodology for any new proposal.

In 1980, the country is 227 million people strong, including 17 million hunters, 42 million anglers, and 62 million people feeding birds (USDI Fish and Wildlife Service and U.S. Bureau of the Census 1982). With most of the country living in metropolitan areas, a new niche is evolving: the urban wildlife biologist. But out on our prairies, eighty thousand waterfowl are dying of avian cholera in Nebraska's Rainwater Basin, victims of too few wetlands (Tiner 1984), and thousands of prairie dogs are dying from poison (Benedict et al. 1996), victims of outmoded federal- and state-management policies for cattle grazing. Yet, our oak lays down good wood in 1980. The Alaska National Interest Lands Conservation Act creates ten new national parks and monuments, doubles the National Wildlife Refuge system, expands our two largest national forests, and adds 56 million acres to our wilderness system.

The good oak reaches twenty-eight feet in 1981, as the Sagebrush Rebellion, an uprising headed by western politicians, ranchers, and mineral developers, tries unsuccessfully to transfer federal lands to state control. That year, the first comprehensive ecological study of the Pacific Northwest rain forest is published (Franklin et al. 1981). The oak grows in 1982, as 2.4 million hunters harvest 50 million mourning doves (Tomlinson et al. 1994) and biologists implicate ring-necked pheasants in the disappearance of Michigan's last prairie chickens (Westemeier and Edwards 1986). The oak's annual ring is wide in 1983. El Niño's warm waters disrupt the food chain for Pacific seabirds, marine mammals, and salmon while, in the southern states, the decimated bottomland hardwood forests of the lower Mississippi floodplain cannot buffer April's massive spring floods (Tiner 1984). In 1984, the oak produces its first acorns as U.S. golf courses, perpetually in need of manicures and chemicals, now number twelve thousand, consume 1.3-million acres (Edmonson 1987), and provide grazing habitat for thousands of Canada geese. That same year, peregrine falcons once again nest east of the Mississippi (Ehrlich et al. 1988), and westward in Montana, the Rocky Mountain Elk Foundation is founded to conserve and enhance wildlife habitats.

In 1985, the oak's annual ring is narrow, reflecting a drought year and the lowest number of continental ducks in forty years (Kresl et al. 1996). In Pennsylvania, Cooperative Farm-Game projects thrive and twenty thousand farmers provide access to the state's hunters (Lagenbach 1986), and the new, federal Conservation Reserve Program funds farmers to set aside croplands for wildlife. The Western Hemisphere Shorebird Reserve Network is launched to protect millions of shorebirds and the wetlands they depend on. The oak is thirty-three feet tall in 1986, when Canada and the United States sign the first North American Waterfowl Management Plan, aimed at restoring populations of waterfowl and protecting their habitat. The oak does not notice, however, the ship from Europe dumping its ballast water laden with zebra mussels into Lake St. Clair or the mussel invasion of Lake Erie two years later, which changes the lake's ecology (Luoma 1996).

In 1987, an El Niño year, the good oak has a bumper crop of acorns. Wildlife biologists capture AC-9, the last wild California condor, for a captive breeding program (Kiff 1990). Congress orders the Forest Service to cut 18% more timber from nine Pacific Northwest forests. The Sierra Club Legal Defense Fund files suit for tweny-five plaintiffs, including the northern spotted owl, challenging the Fish and Wildlife Service's refusal to list the owl under the Endangered Species Act (Caufield 1990). The oak turns forty in 1988, the year of the First Midwest Oak Savanna Conference. It grows slowly amidst yet another drought, but produces no acorns. The summer's savanna burn on the Leopold farm puts a small fire scar on the oak's thick trunk. Westward in Yellowstone, the Hellroaring Fire sweeps through one-third of the park from July until November and sparks a controversy over the "let-burn" policy.

In 1989, the oak does not notice the massive destruction wrought by both nature and humans. It does not hear the trees snap as Hurricane Hugo wreaks havoc on Puerto Rico's rain forests and devastates South Carolina's coastal pine forests, blowing down one billion board feet in one national forest, wiping out most nest trees used by its 475 red-cockaded woodpecker colonies (Graham 1990). It does not hear the mournful calls of dying, oil-soaked seabirds and marine mammals nor notice the millions of dead salmon and herring when a tanker spews eleven million gallons of crude oil into Alaska's coastal waters. The oak does not heed a new report's warning: this country's economic pursuits and need for motorized leisure vehicles are hurting our refuge system's wildlife (Williams 1996).

"Clear it!" cry the developers, and we pause for breath.

The oak enters the 1990s, its growth rings waxing and waning with the vagaries of El Niño, La Niña, and global warming—a natural phenomenon aggravated by the industrial by-products produced to satisfy 249 million American consumers and their fellow 5.7 billion world citizens. In 1990, the oak's height is thirty-six feet when Earth Day, now twenty years old, is celebrated by two hundred million people in 140 countries (Anonymous 1991).

That year a prominent scientist declares, "We've been running an ecological deficit and the bill has come in. There's going to be pain for owls, people, and for trees (Jerry Franklin, quoted in Gup 1990, 60). The federal government lists the northern spotted owl as a threatened species and implements a conservation strategy that includes a system of large tracts of mature and old-growth forests on public lands. Thousands of loggers protest a future without jobs.

In spring 1991, the oak hears the sandhill cranes returning from several weeks along the Platte River with millions of waterfowl and eighty thousand crane watchers (Lingle 1992). The oak does not notice that Massachusetts regulates off-road vehicles on coastal beaches, stemming the devastation and disturbance to both the ecosystem and the endangered piping plovers attempting to rear their young (Scott Melvin, Massachusetts Division of Fish and Wildlife, pers. commun.). Nor does the oak notice forty-nine captive-reared black-footed ferrets scampering into the burrows of a Wyoming prairie dog town that fall (Raloff 1992), marking yet another attempt to restore a nearly lost species to the wild. In 1992, the oak does not heed the complaints and claims when a neighboring state's wolf packs depredate 10 cows, 53 calves, 22 sheep, 526 turkeys, 1 horse, 6 chickens, and 8 dogs, along with an occasional pig, goat, and goose (William Paul, USDA, unpubl. rept.).

In 1993, the oak's annual ring is wide, recording a winter of heavy snows and a warm, wet spring. It barely notices that spring's "storm of the century," a blizzard that buries lands from the Gulf Coast to New England, nor does it hear the thunderclaps as California's seven-year drought finally ends. But in June, its limbs bend low during torrential rains, and in July, it hears the dams burst and the water rush over the levees of the mighty Mississippi as the Great Flood of 1993 covers over sixteen million acres of farmland (Anonymous 1994). The oak grows in 1994, when the Wildlife Society publishes its fifth book in thirty-four years on wildlife techniques (Bookhout 1994) and Congress limits nest protection for Queen Charlotte goshawks to three hundred acres (Iverson et al. 1996). That same year, federal agencies purchase 45,000 acres of California's Merced, San Joaquin, and Stanislaus rivers to aid the migration of fall-run chinook salmon to the Pacific Ocean (Ducks Unlimited 1994). In southern California, two new national parks and a new national preserve are created to provide federal protection to nearly 2.5 million acres of desert.

In 1995, the oak stands forty feet tall. It does not notice when the flock of northern orioles that lit in its bough the previous fall never return, all dead on collision with a new telecommunications tower in their migration pathway. In 1996, the oak does not heed a new book, *Our Stolen Future* (Colborn et al. 1996), whose message carries the most important warning of the decade: synthetic chemicals are mimicking natural hormones and impairing reproduction for a variety of wildlife, including American alligators, Florida panthers, and bald eagles; nor does the oak notice that the booming grounds of Attwater's prairie chicken are all but deserted.

The oak grows six inches in 1997, the earth's warmest year on record. That year, a federal district judge rules the gray wolf recovery program illegal and orders Yellowstone's reintroduced wolves and offspring removed (Anonymous 1998). Waterfowl populations reach their highest levels in twenty-five years, and, after years of study, controversy, and political wrangling, a regional forester signs a new management plan for our nation's largest forest. The plan ensures population viability for key wildlife species using old-growth habitat reserves and beach and estuary buffers (USDA Forest Service 1997).

In 1998, the oak grows as the World Trade Organization rules unfettered trade more important than environmental considerations, such as use of sea turtle excluders (Cushman 1998). The oak groans as one of its lower branches breaks off and crashes during a late spring blizzard. But it pays no heed when, nearby in Madison, one of the world's best examples of a restored prairie is threatened with the encroachment of a housing development (Bill Jordan, University of Wisconsin Arboretum, pers. commun.), or when Florida's pinewoods go up in flames. Nor does the oak pay heed when scientists warn that 29% of the 16,108 plant species in the United States are under threat of extinction (Sadowski and Strahm 1998).

After all, the good oak is safe on the farm. Or is it?

References

Anderson, J. M. 1987. Restoring nongame wildlife. In *Restoring America's wildlife 1937–1987*, ed. H. Kallman, 229–241. Washington, D.C.: USDI Fish and Wildlife Service, U.S. Government Printing Office.

Anonymous. 1991. Our fragile earth. In *The new book of knowledge* (annual), ed. L.T. Lorimer, 114–120. Danbury, Conn.: Grolier Inc.

Anonymous. 1994. The great flood of 1993. In *The new book of knowledge* (annual), ed. L.T. Lorimer, 112–117. Danbury, Conn.: Grolier Inc.

Anonymous. 1998. Wolf recovery program. *The Wildlifer* 286:7.

Belanger, D. O. 1988. *Managing American wildlife: a history of the International Association of Fish and Wildlife Agencies.* Amherst: University of Massachusetts Press.

Bellrose, F. C. 1980. *Ducks, geese, and swans of North America.* Harrisburg, Pa.: Stackpole Books.

Benedict, R. A., P. W. Freman, and H. H. Genoways. 1996. Prairie legacies—mammals. In *Prairie conservation: preserving North America's most endangered ecosystem*, ed. F. B. Samson and F. L. Knopf, 149–167. Washington, D.C.: Island Press.

Bookhout, T. A., ed. 1994. *Research and management techniques for wildlife and habitats.* 5th ed. Bethesda, Md.: The Wildlife Society.

Buckley, J. L., and P. F. Springer. 1964. Insecticides. In *Waterfowl tomorrow*, ed. J. R. Linduska, 459–470. Washington, D.C.: U.S. Government Printing Office.

Burwell, R. W., and L. G. Sugden. 1964. Potholes—going, going . . . In *Waterfowl tomorrow*, ed. J. R. Linduska, 369–380. Washington, D.C.: U.S. Government Printing Office.

Carson, R. 1962. *Silent spring.* Boston, Mass.: Houghton Mifflin.

Caufield, C. 1990. A reporter at large, the ancient forest. *The New Yorker*, 14 May, 46–52.

Coates, P. A. 1991. *The trans-Alaska pipeline controversy: technology, conservation, and the frontier.* Bethlehem, Pa.: Lehigh University Press.

Colborn, T., D. Dumanoski, and J. P. Myers. 1996. *Our stolen future: are we threatening our fertility, intelligence, and survival? A scientific detective story.* New York: Dutton.

Cushman, J. H., Jr. 1998. Trade group strikes blow at U.S. environmental law. *New York Times*, 7 April, C:1, 5.

Ducks Unlimited. 1994. Federal agencies purchase water for San Joaquin Valley salmon. *Valley Care* 1(2):5.

Edmonson, J. 1987. Hazards of the game. *Audubon* 89(6):24–37.

Ehrlich, P. R., D. S. Dobkin, D. Wheye. 1988. *The birder's handbook: a field guide to the natural history of North American birds.* New York: Simon and Schuster.

Franklin, J. F., K. Cromack, Jr., W. Denison, A. McKee, C. Maser, J. Sedell, R.

Swanson, and G. Juday. 1981. Ecological characteristics of old-growth Douglas-fir forests. USDA Forest Service, General Technical Report PNW-118.
Graham, F., Jr. 1987. The "sewer ditch" undone. *Audubon* 89(2):114–115.
Graham, F., Jr. 1990. Matchsticks! *Audubon* 92(l):44–51.
Gup, T. 1990. Owl vs man. *Time* 135(26):56–62.
Hawkins, A. S. 1980. The role of hunting regulations. In *Ducks, geese and swans of North America*, ed. F. C. Bellrose, 59–63. Harrisburg, Pa.: Stackpole Books.
Horn, E. E., and L. L. Glasgow. 1964. Rice and waterfowl. In *Waterfowl tomorrow*, ed. J. P. Linduska, 435–444. Washington, D.C.: U.S. Government Printing Office.
Hornocker, M. G., and H. Quigley. 1987. Mountain lion: Pacific Coast predator. In *Restoring America's wildlife 1937–1987*, ed. H. Kallman, 177–189. Washington, D.C.: USDI Fish and Wildlife Service, U.S. Government Printing Office.
Iverson, G. C., G. D. Hayward, K. Titus, E. DeGayner, R. E. Lowell, D. C. Crocker-Bedford, P. F. Schempf, and J. Lindell. 1996. Conservation assessment for the northern goshawk in southeast Alaska. USDA Forest Service, Pacific Northwest Research Station, General Technical Report PNW-GTR-387.
Kiff, L. 1990. To the brink and back, the battle to save the California condor. *Terra* 28(4):7–18.
Kresl, S. J., J. T. Leach, C. A. Lively, and R. E. Reynolds. 1996. Working partnerships for conserving the nation's prairie pothole ecosystem: the U.S. Prairie Pothole Joint Venture. In *Prairie conservation: preserving North America's most endangered ecosystem*, ed. F. B. Samson and F. L. Knopf, 203–210. Washington, D.C., Island Press.
Lagenbach, J. R. 1986. Restoring a land base. In *Restoring America's wildlife 1937–1987*, ed. H. Kallman, 69–77. Washington, D.C.: USDI Fish and Wildlife Service, U.S. Government Printing Office.
Leopold, A. 1949. *A sand county almanac and sketches here and there*. New York: Oxford University Press.
Lewis, J. B. 1987. Success story: wild turkey. In *Restoring America's wildlife 1937–1987*, ed. H. Kallman, 31–43. Washington, D.C.: USDI Fish and Wildlife Service, U.S. Government Printing Office.
Lingle, G. R. 1992. History and economic impact of crane-watching in central Nebraska. *Proceedings of the North American Crane Workshop* 6:33–37.
Luoma, J. R. 1996. Biography of a lake. *Audubon* 98(5):66–72.
Martin, A. C., H. S. Zim, and A. L. Nelson. 1951. *American wildlife and plants, a guide to wildlife food habits: the use of trees, shrubs, weeds, and herbs by birds and mammals of the United States*. New York: McGraw-Hill Book Company.
Matthiessen, P. 1959. *Wildlife in America*. New York: Viking Press.

McHarg, I. L. 1969. *Design with nature*. Garden City, N.Y.: American Museum of Natural History, Natural History Press.

McNulty, F. 1966. *The whooping crane*. New York: Dutton.

Mosby, H. S. 1960. *Manual of game investigational techniques*. Washington, D.C.: The Wildlife Society.

Rakestraw, L. W. 1981. *A history of the United States Forest Service in Alaska*. Anchorage: Alaska Historical Commission, Department of Education.

Raloff, J. 1992. Environment. In *The Americana annual 1993*, ed. L.T. Lorimer, 226–228. Danbury, Conn: Grolier Inc.

Sadowski, M., and W. Strahm. 1998. First ever IUCN Red List of threatened plants. *Species* 30:22–25.

Scudder, H. C. 1970. The Alaska salmon trap: its evolution, conflict, and consequences. *Historical Monograph No. 1*, Alaska State Library, Juneau.

Stap, D. 1996. Returning the natives. *Audubon* 98(6):5440.

Thiel, R. P. 1993. *The timber wolf in Wisconsin: the death and life of a magnificent predator*. Madison: University of Wisconsin Press.

Thiel, R. P., and R. J. Welch. 1981. Evidence of recent breeding activity in Wisconsin wolves. *American Midland Naturalist* 106:401–402.

Thomas, J. W., ed. 1979. *Wildlife habitats in managed forests—the Blue Mountains of Oregon and Washington*. USDA Forest Service Agriculture Handbook No. 553.

Tiner, R. W. 1984. *Wetlands of the United States: current status and recent trends*. Washington, D.C.: USDI Fish and Wildlife Service.

Tomlinson, R. E., D. D. Dolton, R. R. George, and R. E. Mirarchi. 1994. Mourning dove. In *Migratory shore and upland game bird management in North America*, ed. T. C. Tacha and C. E. Braun, 5–26. Washington, D.C.: International Association of Fish and Wildlife Agencies.

USDA Forest Service. 1996. Tongass land management plan revision: revised supplement to the environmental impact statement. USDA Forest Service, General Report RIO-AIB-314a.

USDA Forest Service. 1997. Tongass Forest plan review. USDA Forest Service, Juneau, Alaska, *Forest Plan Revision Newsletter* 16.

USDI Fish and Wildlife Service. 1983. Endangered and threatened wildlife and plants. U.S. Department of the Interior, Washington, D.C. 50 CFR 17.11 and 17.12.

USDI Fish and Wildlife Service and U.S. Bureau of the Census. 1982. *1980 national survey of fishing, hunting, and wildlife-associated recreation*. Washington, D.C.: U.S. Government Printing Office.

Westemeier, R. L., and W. R. Edwards. 1986. Prairie-chicken: survival in the Midwest. In *Restoring America's wildlife 1937–1987*, ed. H. Kallman, 119–131. Wash-

ington, D.C.: USDI Fish and Wildlife Service, U.S. Government Printing Office.
Williams, T. 1996. Seeking refuge. *Audubon* 98(3):34–45.
Williamson, L. L. 1987. Evolution of a landmark law. In *Restoring America's wildlife 1937–1987*, ed. H. Kallman, 1–29. Washington, D.C.: USDI Fish and Wildlife Service, U.S. Government Printing Office.

The Cohesive Vision

Chapter 6

From the Balance of Nature to the Flux of Nature

The Land Ethic in a Time of Change

J. BAIRD CALLICOTT

Ecology Then and Now

In the fall of 1905, Aldo Leopold was a college freshman in the Sheffield Scientific School at Yale University. That same year, *Research Methods in Ecology*, by Frederic Clements, was published. It was the first textbook in what was then a brand-new scientific discipline. Doubtless, it was required reading for Leopold's training at the Yale Forest School, which he completed in 1909. Clements (1905) believed that a "plant association" or community was a "superorganism." He reasoned as follows. The first organisms were single celled. Over time, some of them evolved symbioses that grew to be so intense that a new, second-order organic unity emerged, the multicelled organism. These organisms, in turn, evolved symbioses to the extent that a third-order organic unity also emerged. These third-order organisms lacked a biological name because they were so diffuse, and they developed so slowly that we intelligent, second-order, multicelled organisms—who are, as it were, moving through them—did not recognize them as organisms. But like ourselves, they *are* organisms, but organisms at the next level up in biological organization. These superorganisms were the proper objects of ecological study according to Clements. Ecology was, thus, understood to be a kind of exophysiology (Semper 1881). The ecologist studied the functions of the various components of superorganisms. As cells are to multicelled organisms, multicelled organisms are to superorganisms; and as organs are to multicelled organisms, species populations are to superorganisms.

Clements specialized in what he conceived to be the developmental biology of superorganisms, plant succession. In 1916, Leopold was in the fifth year of his career with the U.S. Forest Service in the Southwest Territories of Arizona and New Mexico. That same year, Clements's *Plant Succession: An Analysis of the Development of Vegetation* was published. Clements (1916) believed that after an outside disturbance, such as the cutting down of a forest or the plowing of a prairie, the disturbed area, if then left alone, would go through a series—each of which he dubbed a "sere"—of plant associations until a "climax" community was achieved. The climax community was the mature superorganism; the seres leading up to it were its developmental stages. Once the mature climax community became established, it would reproduce itself in perpetuity until again disturbed by some outside force. Clements (1916) believed that in nature undisturbed by human beings, climax-destroying forces were so infrequent as to be abnormal. But when they did occur, succession—if not forcefully arrested—would proceed through the same series of seres until the same climax was reestablished.

In 1935, Aldo Leopold purchased a worn-out Wisconsin River farm with one standing building, a chicken coop, that he and his family converted into the "shack." By then, he had become the first professor of game management in the United States. That same year, "The Use and Abuse of Vegetational Concepts and Terms," by Arthur Tansley was published. In this influential article, Tansley rejected Clements's concept of a superorganism and coined the term "ecosystem" for the proper objects of ecological study. But Tansley's break from Clementsian ecology was far from radical. He wrote:

> The relatively stable climax community is a complex whole with more or less definite structure, i.e., inter-relation of parts adjusted to exist in the given habitat and coexist with one another. It has come into being through a series of stages which have approximated more and more to dynamic equilibrium in these relations. This surely is "organisation," and organisation of the same type as, though by no means identical with, that of the single animal or plant. (Tansley 1935, 291)

For Tansley, the climax community was not a superorganism; rather, it was an "adult quasi-organism." Moreover, ecosystems, as Tansley (1935, 300) conceived them, were evolving entities, subject to natural selection: "There is in fact a kind of natural selection of incipient systems, and those which can attain the most stable equilibrium survive the longest. . . . The universal tendency to the evolution of dynamic equilibria has long been recognized."

In 1942, Aldo Leopold began work on the book that would become *A Sand County Almanac*. That same year, "The Trophic-Dynamic Aspect of Ecology" by Raymond Lindeman was published. In this influential article, Lindeman added an important element missing in Tansley's original characterization of ecosystems: energy. In addition to their physiology (the functions of their component species) and ontogeny (their developmental stages), Lindeman initiated the study of the metabolism (energy processing) of the quasi-organismic ecosystems, which were in a long-evolved condition of dynamic equilibrium. This was the state of the art in ecology when Aldo Leopold wrote "The Land Ethic."

By 1948, Aldo Leopold was dead. The equilibrium ecology of Clements, Tansley, and Lindeman lived on, however, in the most widely used ecology textbook of the mid-twentieth century: *Fundamentals of Ecology* by Eugene Odum, first published in 1953, five years after Leopold's death. That book remained the standard text in the field for the next two decades. The dominant Clementsian ecology had been challenged in the 1920s by H. A. Gleason, but his "individualistic" conception of plant associations had been virtually ignored. Gleason (1926) noted, among other things, the following: the spatial boundaries of plant communities are fuzzy, as are the temporal boundaries between seres; on close inspection, what looks like similar communities of the same type are quite heterogeneous in composition; and successional change is directionless, that is, not tending toward any particular end point or climax. Then he asked rhetorically, "Are we not justified in coming to the general conclusion, far removed from the prevailing opinion, that an association is not an organism, scarcely even a vegetational unit, but merely a *coincidence*?" (Gleason 1926, 16). By the mid-1950s, quantitative empirical analysis of the composition and structure of putative plant communities began to bear out Gleason's suspicions (McIntosh 1998). (Incidentally, among the early neo-Gleasonians was John Curtis, Leopold's colleague at the University of Wisconsin.) Serious doubts about the existence of quasi-organismic levels of biological organization beyond the multicelled organism began to be expressed.

Another blow to what R. H. Whittaker (1967) called the "community unit theory" came from paleoecology—the study of past ecological conditions primarily by dating and examining pollens preserved in bog and lake sediments (Davis 1984). Tansley (1935, 302) vividly captured one corollary of the Clementsian community unit theory: "If a continental ice sheet slowly and continuously advances or recedes over a considerable period of time all the zoned climaxes which are subjected to the decreasing or increasing temperature will, according to Clements's conception, move across the continent 'as if they were strung on a string.'" In an essay written in 1944, "Conservation in Whole or in Part?" Leopold endorses this idea:

> The Wisconsin land was stable . . . for a long period before 1840 [when settled by Europeans]. The pollens embedded in peat bogs show that the native plants comprising the prairie, the hardwood forest, and the coniferous forest are about the same now as they were at the end of the glacial period, 20,000 years ago. Since that time these major plant communities were pushed alternatively northward and southward several times by long climatic cycles, but their membership and organization remained intact. . . . The bones of animals show that the fauna shifted with the flora, but its composition or membership likewise remained intact. (Leopold 1991, 311–312)

Better resolution of the pollen record now reveals a very different picture. Linda Brubaker sketches it:

> Because species have responded individualistically to climatic variations, plant communities have been transient assemblages, seldom persisting more than 2,000 to 5,000 years. . . . Most of the tree species dominating North America today became common 8,000 to 10,000 years ago when they expanded from ice-age refugia. Most species spread at different rates and in different directions, reaching their current range limits and populations only 3,000 to 5,000 years ago. Thus present-day North American forests should not be considered stable over evolutionary time scales. (Brubaker 1988, 41)

Michael Soulé, one of the architects of conservation biology, starkly summarizes this and other equally troubling developments in ecology during the last quarter of the twentieth century:

> Certainly the idea that species live in integrated communities is a myth. So-called biotic communities, a misleading term, are constantly changing in membership. The species occurring in any particular place are rarely convivial neighbors; their coexistence in certain places is better explained by individual physiological tolerances. Though in some cases the finer details of spatial distribution may be influenced by positive interspecies interactions, the much more common kinds of interactions are competition, predation, parasitism, and disease. Most interactions between individuals and species are *selfish*, not symbiotic. Current ecological thinking argues that nature at the level of local biotic assemblages has never been homeostatic. The principle of balance has been replaced with the principle of gradation—a continuum of degrees of . . . disturbance. (Soulé 1995, 143)

Let's face it, two of the most fundamental organizing concepts of twentieth-century ecology, the biotic community concept and the ecosystem concept, are now in doubt (Shrader-Frechette and McCoy 1993). Is a biotic community just a fortuitous assembly of organisms all adapted to similar gradients of moisture, temperature, soil pH, etc.? Just what *are* ecosystems as objects of scientific study? They seem ontologically vague and ambiguously bounded. The very idea that nature is somehow stable—that is, in a dynamic *equilibrium* (such as Tansley imagined) of countervailing forces, like that once believed to exist between a predator and its prey—is passé (Shrader-Frechette and McCoy 1993; Sagoff 1985). Rather, nature is simply dynamic, ever changing (Botkin 1990).It is, moreover, chaotic, changing unpredictably (Degan et al. 1987; Gleik 1987; May 1974). That's all bad enough for the land ethic, but it gets worse. In the 1970s and 1980s, ecologists began to discover that a disturbance or "perturbation" was ecologically "incorporated" and normal, not external and abnormal. Disturbance—by wind, flood, fire, pestilence—not freedom from such disruption, is part of nature's normal state (Pickett and White 1985). Thus, the concept of "disturbance regimes" was introduced to the discipline (Pickett and White 1985).

Famously, in "The Land Ethic," Leopold (1949, 204) wrote, "[A] land ethic changes the role of *Homo sapiens* from conqueror of the land community to plain member and citizen of it. It implies respect for his fellow citizens and also respect for the community as such." And even more famously, he wrote,

"A thing is right when it tends to preserve the integrity, stability, and beauty of the biotic community. It is wrong when it tends otherwise." (Leopold 1949, 204–205). Does it make sense to try to preserve a mere coincidence? Is it even meaningful to talk about the integrity, stability, and beauty of levels of biological organization whose existence is doubtful? And what can we appropriately *preserve*, if nature is directionlessly dynamic? Further, if nature is constantly disturbing biotic communities and ecosystems—allowing, for a moment, that such entities are not relegated to the dustbin of discarded scientific ideas along with phlogiston and the luminiferous ether—what could be wrong with the disturbances *we* impose on it? In short, What implications do the reductive, dynamistic, and disturbing (pun intended) developments in contemporary ecology have for Leopold's land ethic, now more than fifty years after he conceived it? First, I will reassess the Leopold land ethic in face of the challenge to the very existence of biotic communities (and ecosystems) posed by the neo-Gleasonian turn in ecology. Then, I will reassess the Leopold land ethic in regard to the challenge posed to it by the emphasis on directionless change and the incorporation and normalization of disturbance in ecology.

Human and Biotic Communities Compared

No neo-Gleasonian ecological theory of which I am aware asserts that organisms are entirely independent of one another. However individualistic and self-seeking each organism may be, consumers cannot exist without producers and producers cannot exist without decomposers. In trying to diminish the threat to environmental ethics (similar to Leopold's) posed by state-of-the-art theory in contemporary ecology, Donald Worster, in the second edition of *Nature's Economy* (1994, 429), evokes "the principle of interdependency": "No organism or species of organism has any chance of surviving without the aid of others." Though the two-species "daisy world" envisioned by James Lovelock (1988) may be of heuristic value in illustrating how the Earth might have evolved an organic thermostat, a world consisting of only two species of the same genus is not ecologically feasible.

And R. P. McIntosh, himself a leading neo-Gleasonian, offers the following caveat about too readily leaping to the conclusion that there is no ecological order in nature, however complex and disequilibrial it may now appear to be:

> The implication of anarchy, or lack of any order, is a common misrepresentation of Gleason's individualistic concept, which some have erroneously said is a random assemblage of species lacking any relations among the species. Neither Gleason nor any of his successors ever said that. Not *all* things are possible in an individualistic community, only some. The resulting pattern is more elusive than in a purported organismic community, but it is certainly not anarchy or random. . . . It is doubtful . . . that any ecologist envisioned a community as a merely chance aggregation of organisms and environment lacking discernible pattern. Gleason and his successors recognized patterns of gradual change of species composition in space and time, in contrast with the putative patterns of change of integrated groups of organisms. (McIntosh 1998, 431, 433)

Let's make a fresh beginning. The community concept in ecology is a metaphor. The metaphor assimilates the way proximate organisms are mutually dependent to the way proximate human beings are mutually dependent. Now let's consider human communities, the reference objects to which biotic associations are putatively analogous. Human communities—at least recent human communities—are neither stable nor typological. They change over time and, in the process of change, they do not come and go as units.

Take my (now erstwhile) community, Stevens Point, Wisconsin, as an example. It began as a logging community in the 1840s. After the old-growth Wisconsin pinery was leveled, the Scandinavian loggers who cut it down moved on to virgin forests farther West, and the land was settled by Polish dairy farmers. But second-growth woods sprang up here and there and some lumber mills hung on, so that while the pestilential Scandinavian arboreal parasites diminished in number, some remained and adapted to the new colonizers, supplying lumber for houses and barns. Loggers and farmers like to drink beer, so an empty niche opened up and German *Braumeisters* invaded the nascent Stevens Point community and established a brewery. Pretty soon, tavern keepers, butchers, bakers, candlestick makers, grocers, farm implement dealers, hardware merchants, Protestant ministers, Catholic priests, doctors, journalists, lawyers, and eventually college professors all drifted in. Journalists, lawyers, and college professors use lots of paper. The old pinery was spontaneously growing back to scrubby aspen and jack pine, and the Wisconsin River runs through it. In addition to nearby woody fiber, there was a handy supply of water power and waste transport, so dams went up and paper mills and mill workers came to town.

Notice that none of these invasive species—with the problematic exception of ministers on a mission, priests following a vocation, and doctors being faithful to their Hippocratic Oath—was primarily motivated by altruism. Nor did they all move in at once, as a unit. In the formative period, each person came because an economic gradient to which he or she was individually adapted presented itself. Community succession in Stevens Point gradually and imperfectly proceeded from timber mill town to paper mill town. Now, Stevens Point has become quite diverse with its social analog of species populations organized into many economic guilds—paper making, hazard insurance, higher education, retailing, manufacturing, and family farming—being most salient. After the community was established, some people who were born there stayed there. Others left. And many people who now immigrate to Stevens Point will leave (as I did) when a more attractive economic gradient (and, in my case, a more attractive climatic gradient) presents itself elsewhere.

Further, the boundaries of the town expanded with the passage of time. The posted "city limits" are not helpful in demarcating the actual community, which comprises several smaller adjacent municipalities and includes, more or less, hundreds of people living on nearby ten- to forty-acre rural estates and eighty- to one thousand-acre farms. But does all this mean that there is no such thing, no community *as such*, that we can call Stevens Point, Wisconsin, that all that exists is just a standing crop of various types of transient, selfish individuals who happen to be adapted to similar socioeconomic gradients in an ill-defined place?

Try to tell that to Stevens Pointers, a very community-minded lot. Sure, most people in Stevens Point, like most people everywhere, devote most of their time to private gain and leisure pursuits. Economic competition is vigorous. But the people of Stevens Point nevertheless respect the fellow members of their municipal community: no one is homeless or hungry; every kid who wants to can join the YMCA, whether or not his or her parents can pay the dues (I supported one or two such memberships myself); and the developmentally challenged are well integrated into the social mainstream. Stevens

Pointers also have respect for the community as such: people root for the home sports teams; a local ordinance prohibiting fringe shopping malls was passed to preserve main street (though, I'm sorry to say, Stevens Point still has tacky strip development); and people impose relatively high property taxes on themselves to support good schools and well-maintained roads, public buildings, and municipal parks.

My point is that contemporary human communities are no more integrated, nor less dynamic, nor any easier to demarcate than biotic communities as represented in neo-Gleasonian ecology. Yet, human communities such as Stevens Point are still recognizable entities and engender moral duties and obligations both to fellow members and to the community as such. On this crucial point, my conclusions, therefore, are these. First, although biotic communities are not now conceived to be so unified and persistent as they once were, they do exist. Proximate plants and animals, however competitive, predatory, and parasitic, are no more independent of one another than are proximate persons. And second, if paradigmatic human communities are sufficiently robust to engender civic duties and obligations both to fellow members and to the communities, then biotic communities, which are not less robust than paradigmatic human communities, are, by parity of reasoning, also sufficiently robust to engender analogous environmental duties and obligations.

The Land Ethic Dynamized

But *what* environmental duties and obligations does human membership in biotic communities generate? To preserve the integrity, stability, and beauty of the biotic community? These cardinal values of the Leopold land ethic may have to be revised—"dynamized," to coin a word—if they are to be ecologically credible. In "The Land Pyramid" section of "The Land Ethic," surprisingly (perhaps), Leopold (1949, 214) expressly rejects the balance-of-nature idea and embraces natural change. He writes, "The image commonly employed in conservation education is 'the balance of nature.' For reasons too lengthy to detail here, this figure of speech fails to describe accurately what little we know about the land mechanism." In "A Biotic View of Land," Leopold does detail his misgivings about the balance-of-nature metaphor:

> To the lay mind, balance of nature probably conveys an actual image of the familiar weighing scale. There may even be danger that the layman imputes to the biota properties which exist only on the grocer's counter. To the ecological mind, balance of nature has merits and also defects. Its merits are that it conceives of a collective total, that it imputes some utility to all species, and that it implies oscillations when balance is disturbed. Its defects are that there is only one point at which balance occurs and that balance is normally static. (Leopold 1939, 728)

As to embracing change, Leopold (1949, 216–217) goes on in "The Land Pyramid" section to emphasize nature's inherent dynamism: "When a change occurs in one part of the circuit," he writes, "many other parts must adjust themselves to it. Change does not necessarily obstruct or divert the flow of energy [through ecosystems]; evolution is a long series of self-induced changes, the net result of which has been to elaborate the flow mechanism and to lengthen the circuit." Leopold knew that conservation must aim at a moving target. How can we conserve a biota that is dynamic, ever changing, when the very words "conserve" and "preserve"—especially when linked to "integrity" and "stability"—connote arresting change? The key to solving that conundrum is the concept of scale. Scale is a general ecological concept that includes rate, as well as scope; that is, the concept of scale is both temporal and spatial. And a review of Leopold's "The Land Ethic" reveals that he had the key, though he may not have been aware of just how multiscalar change in nature actually is.

Also in "The Land Pyramid" section, Leopold (1949, 217) writes, "Evolutionary changes . . . are usually slow and local. Man's invention of tools has enabled him to make changes of unprecedented violence, rapidity, and scope." Leopold was keenly aware that nature is dynamic, but, under the sway of midcentury equilibrium ecology, he conceived of natural change primarily in evolutionary, not in ecological, terms. Nevertheless, scale is equally normative when ecological change is added to evolutionary change, that is, when normal climatic oscillations and patch dynamics are added to normal rates of extinction, hybridization, and speciation.

The scale notion is currently being employed to refine the ecosystem concept in ecology. As noted, a major problem with the ecosystem concept is bounding ecosystems. The field-defining paper by Lindeman (1942) focused on Cedar Bog Lake in Minnesota. The influential work of Herbert Bormann and Gene Likens (1979) focused on Hubbard Brook. Thus, one way that ecosystems came to be defined was to regard them as coextensive with watersheds (Golley 1994). But such a method of defining ecosystems is crude at best and inapplicable at worst to marine ecosystems and to other study areas that are not easily divisible into watersheds. The watershed method of defining ecosystems is also inapplicable to transwatershed problems such as demarcating the Greater Yellowstone Ecosystem or determining the ecosystemic needs and functions of wide-ranging species like bears and wolves.

However, with the development of hierarchy theory in ecology, ecosystems may be defined quite precisely—albeit both abstractly and relativistically—in reference to temporally scaled processes. According to Tim Allen and Tom Hoekstra:

> Ecosystems are not readily defined by spatial criteria. Ecosystems are more easily conceived as a set of interlinked, differently scaled processes that may be diffuse in space, but easily defined in turnover time. . . . Thus a single ecosystem is itself a hierarchy of differently scaled processes. . . . There are differently scaled processes inside a single ecosystem, as well as sets of differently scaled, more inclusive and less inclusive ecosystems. . . . The degree to which processes of different types express themselves and the length of time they do so, are both ways of describing the uniqueness of particular ecosystems. Much of what we observe in ecosystems is better set in time rather than space. . . . The ecosystem is a much richer concept than just some meteorology, soil, and animals, tacked onto patches of vegetation. . . . Ecosystems can be seen more powerfully as sequences of events rather than things in a place. These events are transformations of matter and energy that occur as the ecosystem does its work. Ecosystems are process-oriented and more easily seen as temporally rather than spatially ordered. (Allen and Hoekstra 1992, 98–100)

Homo sapiens is a part of nature, "a plain member and citizen" of the "land-community," as Leopold (1949, 204) puts this evolutionary-ecological point. Hence, anthropogenic changes imposed on nature are no less natural than any other changes. But, because *Homo sapiens* is a moral species capable of ethical deliberation and conscientious choice, and evolutionary kinship and biotic community membership add a land ethic to our familiar social ethics, anthropogenic changes may be land-ethically evaluated. But by what norm? The norm of appropriate scale.

Let me first elaborate Leopold's use of the temporal scale of evolutionary change as a norm for evaluating anthropogenic change. Consider the current episode of abrupt, anthropogenic, mass species extinction, which many people, including me, intuitively regard as the most morally reprehensible environmental thing going on today. Episodes of mass extinction have occurred in the past, though none of those has been attributed to a biological agent (Raup 1991; Raup and Sepkoski 1984). Such events are, however, abnormal. Normally, speciation outpaces extinction, which is why biodiversity has increased over time. So, what is land-ethically wrong with current anthropogenic species extinction? Species extinction is not unnatural. On the contrary, species extinction—anthropogenic or otherwise—is perfectly natural. But the current *rate* of extinction is wildly abnormal. Does being the first biological agent of a geologically significant mass extinction event in the 3.5-billion-year tenure of life on planet Earth morally become us *Homo sapiens*? Doesn't that make a mockery of the self-congratulatory species epithet: the sapient, the wise species of the genus *Homo*?

Earth's climate has warmed up and cooled off in the past. So what's land-ethically wrong with the present episode of anthropogenic global warming? We are a part of nature, so our recent habit of recycling sequestered carbon may be biologically unique, but it is not unnatural. A land-ethical evaluation of the current episode of anthropogenic climate change can, however, be made on the basis of temporal scale and magnitude. We may be causing a big increase of temperature at an unprecedented rate (Schneider 1989).That's what's land-ethically wrong with anthropogenic global warming.

Temporal and spatial scale in combination are key to the evaluation of direct human ecological impact. Violent nonanthropogenic perturbations regularly occur (Pickett and White 1985). Volcanoes bury the biota of whole mountains with lava and ash. Tornadoes rip through forests, leveling trees. Hurricanes erode beaches. Wildfires sweep through forests, as well as savannas. Rivers drown floodplains. Droughts dry up lakes and streams. Why, therefore, are anthropogenic clear-cuts, beach developments, hydroelectric impoundments, and the like environmentally unethical? As such, they are not. Once again, it's a question of scale. In general, frequent, intense disturbances, such as tornadoes, occur at small, widely distributed spatial scales.

And spatially broadcast disturbances, such as droughts, occur less frequently. And most disturbances at whatever level of intensity and scale are stochastic (random) and chaotic (unpredictable). The problem with anthropogenic perturbations—such as industrial forestry and agriculture, exurban development, drift net fishing—is that they are far more frequent, widespread, and regularly occurring than nonanthropogenic perturbations.

Stewart Pickett and Richard Ostfeld—exponents of the new natural disturbance/patch dynamics paradigm in ecology, which they dub "the flux of nature" (in contrast to the old "balance of nature")—agree that appropriate scale is the operative norm for ethically appraising anthropogenic ecological perturbations. They note that "the flux of nature is a dangerous metaphor. The metaphor and the underlying ecological paradigm may suggest to the thoughtless and greedy that since flux is a fundamental part of the natural world, any human-caused flux is justifiable. Such an inference is wrong because the flux in the natural world has severe limits. . . . Two characteristics of human-induced flux would suggest that it would be excessive: fast rate and large spatial extent" (Pickett and Ostfeld 1995, 273–274).

Among the abnormally frequent and widespread anthropogenic perturbations that Leopold (1949, 217) himself censures in "The Land Ethic" are the continent-wide elimination of large predators from biotic communities, the ubiquitous substitution of domestic species for wild ones, the ecological homogenization of the planet resulting from the anthropogenic "world-wide pooling of faunas and floras," the ubiquitous "polluting of waters or obstructing them with dams."

The Upshot

So let me return, by way of summary and conclusion, to the question that is implicit in the title of this essay: Does the paradigm shift in ecology from the balance of nature to the flux of nature undermine the land ethic?

The answer is No.

Biotic communities may be ever-changing assemblages of organisms of various species that happen to be adapted to the same edaphic and climatic gradients (Whittaker 1967). But that makes them even more analogous to human communities than the old static-holistic representation. Ever-changing, imprecisely bounded communities of human individualists are robust enough to be identifiable entities and to generate special obligations to fellow members and to such communities per se. Why should a communitarian environmental ethic such as Leopold's have to meet any higher standard of community robustness?

The summary moral maxim of the land ethic, however, must be dynamized in light of developments in ecology since the mid-twentieth century. Leopold acknowledged the existence and land-ethical significance of natural environmental change, but he seems to have thought of it primarily on a very slow, evolutionary temporal scale. But even so, he thereby incorporates the concept of inherent environmental change and the crucial norm of scale into the land ethic. In light of more recent developments in ecology, we can add norms of scale to the land ethic for both climatic and ecological dynamics in land-ethically evaluating anthropogenic changes in nature. One hesitates to edit Leopold's elegant prose, but as a stab at formulating a dynamized summary moral maxim for the land ethic, I will hazard the following: *A thing is right when it tends to disturb the biotic community only at normal spatial and temporal scales. It is wrong when it tends otherwise.*

Literature Cited

Allen, T. F. H., and T. W. Hoekstra. 1992. *Toward a unified ecology.* New York: Columbia University Press.

Bormann, F. H., and G. E. Likens. 1979. *Pattern and process in a forested ecosystem.* New York: Springer-Verlag.

Botkin, D. 1990. *Discordant harmonies: a new ecology for the twenty-first century.* New York: Oxford University Press.

Brubaker, L. B. 1988. Vegetation history and anticipating future vegetation change. In *Ecosystem management for parks and wilderness*, ed. J. K. Agee, 42–58. Seattle: University of Washington Press.

Clements, F. E. 1905. *Research methods in ecology.* Lincoln, Nebr.: University Publishing Company.

Clements, F. E. 1916. *Plant succession: an analysis of the development of vegetation.* Publication No. 242. Washington: Carnegie Institution.

Davis, M. B. 1984. Climatic instability, time lags, and community disequilibrium. In *Community ecology*, ed. J. Diamond and T. J. Case, 269–284. New York: Harper and Row.

Degan, H., A. V. Holden, and L. F. Olsen. 1987. *Chaos in biological systems.* New York: Plenum Press.

Gleason, H. A. 1926. The individualistic concept of the plant association. *Bulletin of the Torey Botanical Club* 53:1–20.

Gleik, J. 1987. *Chaos: the making of a new science.* New York: Viking.
Golley, F. B. 1994. *A history of the ecosystem concept in ecology.* New Haven, Conn.: Yale University Press.
Leopold, A. 1939. A biotic view of land. *Journal of Forestry* 37:727–730.
Leopold, A. 1949. *A sand county almanac and sketches here and there.* New York: Oxford University Press.
Leopold, A. 1991. Conservation in whole or in part? In *The River of the Mother of God and other essays by Aldo Leopold,* ed. S. L. Flader and J. B. Callicott, 310–319. Madison: University of Wisconsin Press.
Lindeman, R. L. 1942. The trophic-dynamic aspect of ecology. *Ecology* 23:399–418.
Lovelock, J. 1988. *The ages of Gaia: a biography of our living earth.* New York: W. W. Norton.
May, R. M. 1974. Biological populations with nonoverlapping generations: stable points, stable cycles, and chaos. *Science* 186:645–647.
McIntosh, R. P. 1998. The myth of community as organism. *Perspectives in Biology and Medicine* 41:426–438.
Odum, E. P. 1953. *Fundamentals of ecology.* Philadelphia, Pa.: Saunders.
Pickett, S. T. A., and R. S. Ostfeld. 1995. The shifting paradigm in ecology. In *A new century for natural resources management,* ed. R. L. Knight and S. F. Bates, 261–278. Washington, D.C.: Island Press.
Pickett, S. T. A., and P. S. White. 1985. *The ecology of natural disturbance and patch dynamics.* Orlando, Fla.: Academic Press.
Raup, D. M. 1991. *Extinction: bad genes or bad luck?* New York: W. W. Norton.
Raup, D. M., and J. J. Sepkoski. 1984. Periodicity of extinctions in the geologic past. *Proceedings of the National Academy of Sciences USA* 81:801–805.
Sagoff, M. 1985. Fact and value in ecological science. *Environmental Ethics* 7:99–116.
Schneider, S. H. 1989. *Global warming: are we entering the greenhouse century?* San Francisco, Calif.: Sierra Club Books.
Semper, K. 1881. *Animal life as affected by the natural conditions of existence.* New York: Appleton.
Shrader-Frechette, K., and E. D. McCoy. 1993. *Method in ecology: strategies for conservation.* Cambridge: Cambridge University Press.
Soulé, M. E. 1995. The social siege of nature. In *Reinventing nature? Responses to postmodern deconstruction,* ed. M. E. Soulé and G. Lease, 137–170. Washington, D.C.: Island Press.
Tansley, A. G. 1935. The use and abuse of vegetational concepts and terms. *Ecology* 16:284–307.
Whittaker, R. H. 1967. Gradient analysis of vegetation. *Biological Review* 42:207–264.
Worster, D. 1994. *Nature's economy: the roots of ecology.* 2d ed. Garden City, N.Y.: Anchor Books.

Chapter 7

Aldo Leopold Was a Conservation Biologist

REED NOSS

I was about halfway through my undergraduate education when I first read *A Sand County Almanac*. The book did not have the influence on me then that it should have. Aldo Leopold's writing did not enthrall me as much as my favorite authors in those years—Thoreau, Faulkner, Steinbeck, Hemingway, Abbey, Tolkien. Not a hunter myself—nor an antihunter—I found the frequent references to hunting experiences in the first half of the book distracting rather than engaging. In some ways, the book seemed quaint.

I grew up with natural history as my greatest love; however, with the noteworthy exception of insect collections in the sixth and tenth grades, my formal scientific education prior to college was divorced from the natural world. Pickled specimens, dense textbooks, too much math, and uninspired teachers turned me off from biology. I'd just as soon skip school and hang out in the woods—which I did quite regularly. My impression of biology did not change much in my first year or two of college biology. Aldo Leopold may have had similar experiences. He commented in one of his *Round River* essays, "[T]he average college student who inclines toward natural-history avocations is rebuffed rather than encouraged. . . . Instead of being taught to see his native countryside with appreciation and intelligence, he is taught to carve cats" (Leopold 1953, 193).

My own disenchantment with scholastic biology precipitated a wander through a series of undergraduate majors that eventually led to environmental education. It was through my environmental-education mentors, not biologists, that I first learned of Aldo Leopold. And although my first reading of *A Sand County Almanac* lacked context, later readings in graduate school, after I had returned to biology, influenced me greatly. My early science teachers had hammered into me that science is value free, nonemotional, amoral—just plain facts. They also stressed that real science is indoor, laboratory science—white lab smocks, test tubes, lots of equations, and worst of all, vivisecting rats and pounding nails through turtles' heads. In rereading *A Sand County Almanac*, I realized that scientists could indeed be naturalists first and foremost. They could talk about such things as beauty, love, and right and wrong. I had believed for a long time that studying natural history offered priceless insights on how to live in harmony with the natural world, but my teachers had nearly convinced me that no relationship exists between fact and value. It took a rereading or two of Leopold to teach me that natural science can help us learn how to live.

Looking back, I realize that Aldo Leopold, as much as anyone, influenced me to pursue a graduate education in wildlife ecology and, ultimately, become a conservation biologist. In this essay, I pose some questions about Aldo Leopold in relation to modern conservation biology and its issues. In particular, I ask: (1) Was Aldo Leopold a conservation biologist? (2) What would Leopold say about current issues in land management and conservation? and (3) Is conservation biology the integrating discipline that Leopold hoped for? My answers to these questions are tentative, reflecting my ongoing education as to what is effective conservation in practice.

Was Aldo Leopold a Conservation Biologist?

Aldo Leopold is a parental figure to wildlife biologists and managers, as well as to foresters, and he is as admired by these professionals today as he was in his lifetime. Leopold, however, was always a few steps ahead of his peers and, in many ways, outside the mainstream of the wildlife field he founded. Although much of his career was management oriented and focused on game animals, Leopold was well respected by academic ecologists and eventually came to be considered part of their community. His early studies of landscape change in the southwestern states during his tenure with the U.S. Forest Service contained brilliant ecological observations (Meine 1988). Leopold later became a personal friend of the great British ecologist Charles Elton, who, among others such as the plant ecologist John Weaver, influenced his thinking. In 1947, Leopold was elected president of the Ecological Society of America despite the fact that he was not an active member and seldom attended the society's annual meetings (Meine 1988). The fundamental ecological principles of Leopold's time, such as the food pyramid (which Leopold called the "land pyramid") and the relationship between diversity and stability, were well represented in *A Sand County Almanac.* Many of the essays in the book relate to the intricate interactions among landform, soil, water, vegetation, animals, and other components of ecosystems. Few professionals in either the wildlife or the ecology establishments, then or now, possess the breadth of knowledge Leopold displayed throughout his work. This ability to see larger patterns and processes in nature is the hallmark of a great ecologist.

Can conservation biologists also claim Leopold as a father figure? The answer depends on one's definition of conservation biology. I define it as science in the service of conservation, with conservation defined broadly to include the protection and restoration of native biodiversity and ecological integrity. Those who first used the term "conservation biology" probably did not have quite such a broad definition in mind. The first known written appearance of the term was in the first sentence and lead article of the inaugural issue of the *Journal of Wildlife Management* (Errington and Hamerstrom 1937). Errington and Hamerstrom (1937, 3) began their paper on ring-necked pheasant nest failures with these words: "In the new and growing field of conservation biology. . . ." In those days, wildlife conservation was predominantly utilitarian (analogous to the agricultural utilitarianism of the field of soil conservation) and concerned with species of recreational interest to people. Indeed, Leopold defined game management as "the art of making land produce sustained annual crops of wild game for recreational use" (Leopold 1933, 1). This preoccupation with game species has not changed all that much. The journals of The Wildlife Society, for example, have retained a relatively narrow focus, with "deer and ducks" prominent among the taxa featured in articles, even in recent years, in stark contrast to the journals of the Ecological Society of America and the Society for Conservation Biology (Jensen and Krausman 1993; Bunnell and Dupuis 1995).

Recognition of the global extinction crisis spurred the emergence of a broader conservation biology in the late 1970s and early 1980s. The leaders of this newly defined "metadiscipline" (Jacobson 1990) were clearly interested in the persistence of all species, not just game animals. As Thomas Lovejoy wrote in the foreword to the first edited volume on conservation biology (Soulé and Wilcox 1980a), "[R]eduction in the biological diversity of our planet is the most basic issue of our time" (Lovejoy 1980, ix). Among the contributors to this book were "botanists, zoologists, ecologists, geneticists, evolutionists, a statistician, a mathematician-demographer, a cytologist, a biochemist, an endocrinologist, a sociobiologist and experts in the field of natural resources" (Soulé and Wilcox 1980b, 1). Hence, unlike conventional approaches to wildlife management but clearly in step with Leopold's diverse personal and professional interests, modern conservation biology is richly interdisciplinary. In the 1980s and 1990s, the interests of conservation biologists expanded quickly beyond extinction to encompass many other kinds of biotic impoverishment, including loss of distinct genotypes and ecosystems and potential ways of halting or reversing these losses.

Leopold was concerned about human-caused extinction, though the extent of species depletion known in his day was a far cry from recent estimates. His distress was based not only on ecological and economic concerns, but also on moral grounds: the immorality of ending another life-form's existence, especially given our evolutionary knowledge that all species are related. Leopold's moral outrage was expressed most poignantly in his *Sand County Almanac* essay "On a Monument to a Pigeon," where he noted that "it is a century now since Darwin gave us the first glimpse of the origin of species. We know now what was unknown to all the preceding caravan of generations: that men are only fellow-voyagers with other creatures in the odyssey of evolution. This new knowledge should have given us, by this time, a sense of kinship with fellow-creatures; a wish to live and let live" (Leopold 1949, 109). This passage is one of Leopold's most effective in demonstrating how the perspective gained by biological knowledge influences—or should influence—our values and ethics. The same Darwinian perspective and sense of moral outrage motivates the new generation of conservation biologists today.

How much, during his career, did Aldo Leopold practice or advocate science in the service of conservation? In a general sense, Leopold's career as a whole was marked by the steady and insightful incorporation of ecological principles into conservation philosophy and practice. In the more limited sense of conservation as forestry or game management, Leopold was ahead of his time in applying scientific principles to management, as is well illustrated in his textbook (Leopold 1933). He was a cofounder of The Wilderness Society and unique among wilderness advocates of his day in emphasizing the scientific and ecological values of wilderness (Leopold 1941; Meine 1988). His 1938 report on management of the Huron Mountain Club in the Upper Peninsula of Michigan included a reserve design with a large, wilderness core area to be preserved in its natural condition and a buffer zone where selective cutting and wildlife management were to be encouraged. For this property, Leopold advocated protection of wolves from persecution because the wolves, he noted, "make the Huron Mountain property more unique and valuable than deer possibly can" (Leopold 1938, cited in Meine 1988, 386). In an important joint meeting of the Ecological Society of America and the Society of American Foresters in 1939, Leopold presented an address on his "biotic view of land," wherein he remarked that "the old categories" of useful and harmful species are purely conditional and that "the only sure conclusion is that the biota as a whole is useful" (Leopold 1939, cited in Meine 1988, 393–394).

Also during the 1930s, Leopold was active in committees of the Ecological Society of America, chaired by Victor Shelford, which proposed to represent all of the major ecosystems of North America in a continent-wide network of protected core areas and surrounding buffer zones. This work, which seemed to have little impact during Leopold's or Shelford's lifetime, nevertheless set the stage for such efforts as the United Nations Educational, Scientific, and Cultural Organization's Man and the Biosphere Program (UNESCO 1974) and the Wildlands Project (Foreman et al. 1992; Noss 1992), which have enlisted the efforts of many conservation biologists. Leopold almost certainly would have supported the Wildlands Project's objective to achieve broad-scale connectivity for wide-ranging animals. Not only did he vigorously champion protection and recovery of large carnivores, but, characteristically ahead of his time, he observed that "many animal species, for reasons unknown, do not seem to thrive as detached islands of population" (Leopold 1949, 253).

Aldo Leopold was, by my definition, a conservation biologist. And, with the breadth and depth of his knowledge, his naturalist's understanding of the landscape, and his honest idealism coupled with practicality, he was one whom modern conservation biologists would do well to emulate.

What Would Leopold Say about Current Issues in Land Management and Conservation?

It is tempting to speculate on what an admired historical figure would say about the hot topics of today. After all, who can prove your speculation wrong? In this spirit, I ask, does Leopold's wisdom offer us help in addressing current problems in conservation? From Meine's (1988) excellent biography, we learn that the issues conservation biologists grapple with today are, in many cases, the same general issues that Leopold was intimately involved in more than a half-century earlier. This is disconcerting, in that it suggests that we haven't learned all that much in the intervening years. On the other hand, Leopold's insights might serve us still in grappling with today's conservation issues.

One topic about which Leopold undoubtedly would have much to say is ecosystem management. Leopold likely would support the aim of ecosystem management to focus on the landscape as a whole, rather than on individual components one by one. His essays in *A Sand County Almanac* are replete with such advice. On the other hand, Leopold was a strong defender of wilderness; he emphasized the role of protected areas as habitat for species and as benchmarks for land management experiments. Hence, he likely would not approve of those proponents of ecosystem management who downplay the role of protected areas. An example of this is the Interior Columbia Basin Ecosystem Management Plan, led by the U.S. Forest Service and covering some 144 million acres, half of which is federal land where the plan would have direct authority (Epatko 1998). In its most recent draft, this plan lacks any recommendations for new protected areas and instead proposes minor changes in management practices. To Leopold, managing the land as an "integrated whole" would surely mean recognizing the fundamental historic, scientific, and conservation values of the wild remnants within the whole.

The most important lesson Leopold can teach us about ecosystem management has to do with humility and caution. Many ecosystem management projects make the arrogant assumption that scientists understand ecosystems well enough to manipulate them for long-term commodity production and other uses without losing biodiversity and ecosystem function (Noss and Cooperrider 1994; Stanley 1995). Leopold, on the other hand, was quick to point out the limitations of our knowledge: "The ordinary citizen today assumes that science knows what makes the community clock tick; the scientist is equally sure that he does not. He knows that the biotic mechanism is so complex that its workings may never be fully understood" (Leopold 1949, 220). The best means of understanding, Leopold felt, was through extensive contact with land that is essentially unmanaged: "A science of land health needs, first of all, a base datum of normality, a picture of how healthy land maintains itself as an organism" (Leopold 1949, 251). Although the organism analogy is out of vogue in ecology, the need for reference sites and benchmarks for ecosystem-level experiments remains well appreciated by most scientists. Leopold would have little patience with the current academic fad of wilderness bashing. "It is only the scholar who understands why the raw wilderness gives definition and meaning to the human enterprise" (Leopold 1949, 256).

Another modern trend about which we might seek guidance from Leopold is the gradual shift from commodity production to recreation as a focus of public land management. The 27 April 1998 issue of *High Country News* features several articles documenting this change. Logging on national forest land, for example, is down from twelve billion board feet annually a decade ago to under four million board feet today, and oil and gas activity has dropped by more than 60% (Chilson 1998). At first glance, this shift seems quite positive: with less extraction, there ought to be more room for wilderness to accommodate species sensitive to human disturbances. The increased recreational use that has partially supplanted commodity production on public lands, however, is of the high-intensity, motorized kind (a $400 billion-a-year industry) (Margolis 1998). Available evidence suggests that recreation, especially involving off-road vehicles, could be just as damaging to natural ecosystems as commodity production (Knight and Gutzwiller 1995), yet the mainstream environmental groups have virtually ignored this issue (Margolis 1998).

Leopold would not be gladdened by the booming public-lands recreation industry (Knight 1999). *A Sand County Almanac* is filled with warnings about the dangers of motorized recreation: "Recreation is valuable in proportion to the intensity of its experiences, and the degree to which it differs from and contrasts with workaday life. . . . Mechanized recreation has already seized nine-tenths of the woods and mountains; a decent respect for minorities should dedicate the other tenth to wilderness" (Leopold 1949, 249); "mass-use involves a direct dilution of the opportunity for solitude . . . accommodations for the crowd are not developing (in the sense of adding or creating) anything. On the contrary, they are merely water poured into the already-thin soup" (Leopold 1949, 264–265); and the very last sentence in the book: "Recreational development is a job not of building roads into lovely country, but of building receptivity into the still unlovely human mind" (Leopold 1949, 269). If we were to follow Leopold's wisdom, we would close many roads on public lands and prohibit use of these lands by off-road vehicles of any kind.

Is Conservation Biology the Integrating Discipline That Leopold Hoped For?

I'm not sure about the answer to this question, but I am hopeful. Some things are certain: the traditional natural resource disciplines, with rare exceptions, remain narrowly oriented toward resources of direct human use, display a short-term perspective, seem arrogantly confident in the ability of humans to manage nature wisely, emphasize control and domination over nature, are enamored with high-tech approaches to management, are comfortable with high risk to ecosystems, suffer from disciplinary fragmentation and competition (e.g., wildlife vs. range vs. fisheries vs. forestry), and are still highly responsive to industry and bureaucracies (Noss and Cooperrider 1994). These things were even more true in Leopold's day, and it troubled him deeply. We have made some progress since then, but arguably not enough. Those individuals in the natural resource departments and agencies who bravely challenge the status quo and attempt to break down traditional barriers need all the support they can get.

The short-term, utilitarian ideology of the natural resource disciplines is especially troubling. Leopold was not satisfied with arguments based solely on utility. An avid hunter, he still had to confess that "the woodcock is a living refutation of the theory that the utility of a game bird is to serve as a target, or to pose gracefully on a slice of toast" (Leopold 1966, 334) and that "an October gentian, dusted with tamarack gold, is worth a full stop and a long look, even when the dog signals grouse ahead" (Leopold 1966, 57). He appreciated his woodlot on his sand county farm as much for the diseases and dead trees as for the healthy, living trees. In *Round River*, he presented one of his often-quoted criticisms of the narrow, utilitarian attitude toward nature: "The last word in ignorance is the man who says of an animal or plant: 'What good is it?' If the land mechanism as a whole is good, then every part is good, whether we understand it or not. If the biota, in the course of eons, has built something we like but do not understand, then who but a fool would discard seemingly useless parts? To keep every cog and wheel is the first precaution of intelligent tinkering" (Leopold 1953, 177). This passage might be taken as a plea for enlightened or long-term utilitarianism, and indeed it is, in part. But Leopold's writing as a whole, especially over the last decade and a half of his life, displays his deep appreciation of the value of nature for its own sake. As he wrote in "The Land Ethic," "[I]t is inconceivable to me that an ethical relation to land can exist without love, respect, and admiration for land, and a high regard for its value. By value, I of course mean something far broader than mere economic value; I mean value in the philosophical sense" (Leopold 1966, 239).

Leopold clearly yearned for a discipline more integrated and holistic than the traditional resource fields. He detested disciplinary fragmentation in universities and elsewhere: "There are men charged with the duty of examining the construction of the plants, animals, and soils which are the instruments of the great orchestra. These men are called professors. Each selects one instrument and spends his life taking it apart and describing its strings and sounding boards. This process of dismemberment is called research" (Leopold 1949, 153). "Education, I fear, is learning to see one thing by going blind to another" (Leopold 1949, 158). The discipline that Leopold increasingly looked toward for holistic guidance was the science of ecology, indeed the natural discipline for transcending boundaries. Leopold described ecology as the "fusion point for all the natural sciences" (Leopold 1939; Curt Meine, Intl. Crane Foundation, pers. commun.). Unfortunately, ecology has become more highly specialized and fragmented since Leopold's time. One can easily see this by looking through the tables of contents of the journal *Ecology* over the last few decades. Papers on diversity and communities and ecosystems as wholes have been largely replaced by narrow reports on single species, specific interactions, and biogeochemistry. In many ways, this change reflects the maturation of the field and a commendable increase in knowledge, but it has its downside—namely, a lack of integration. One drowns in a sea of particular facts. Moreover, ecology remains, for the most part, rigidly academic, with applied issues and workers given lower status in the pecking order.

Is conservation biology more integrated than these other fields? Maybe. Conservation biology at least strives to be interdisciplinary, although it is clearly dominated by biologists and especially by ecologists and geneticists. Nevertheless, in comparison with the natural resource disciplines, conservation biology emphasizes a long-term perspective, focuses on several levels of biological organization (e.g., genes through ecosystems), displays more humility about management, seeks to minimize risk to natural ecosystems, seeks cooperation among disciplines and agencies, and is more responsive to the needs of native species and local peoples than to bureaucracies and industry (Noss and Cooperrider 1994). Conservation biology appears to be more demographically diverse than some professions, with many age groups and both men and women well represented (although it is still dominated racially by whites). Compared to ecology, conservation biology is more action oriented and more often assembles interdisciplinary teams to tackle complex planning and management problems.

Science and society remain far from realizing Leopold's vision. Leopold understood that those who appreciate nature for its own sake, who seek out blank spots on the map and would rather see geese than television, are a tiny minority of citizens. Many follow their convictions and become conservationists; some become conservation biologists. We are still a small minority, but we are at least slowing the tide of biotic impoverishment. Or so we hope. We owe it to Leopold to do no less.

References

Bunnell, F. L., and L. A. Dupuis. 1995. Conservation biology's literature revisited: wine or vinaigrette? *Wildlife Society Bulletin* 23:56–62.

Chilson, P. 1998. An era ends. Old industries face reality. *High Country News* 30(8):12–13.

Epatko, L. 1998. Columbia basin management plan on the ropes. *Land Letter* 17(14):1–5.

Errington, P. L., and F. N. Hamerstrom. 1937. The evaluation of nesting losses and juvenile mortality of the ring-necked pheasant. *Journal of Wildlife Management* 1:3–20.

Foreman, D., J. Davis, D. Johns, R. Noss, and M. Soulé. 1992. The Wildlands Project mission statement. *Wild Earth* (Special Issue):3–4.

Jacobson, S. K. 1990. Graduate education in conservation biology. *Conservation Biology* 4:431–440.

Jensen, M. N., and P. R. Krausman. 1993. Conservation biology's literature: new wine or just a new bottle? *Wildlife Society Bulletin* 21:199–203.

Knight, R. L. 1999. Leopold and the "Still Unlovely Mind." In *The essential Aldo Leopold: quotations and commentaries*, ed. C. Meine and R. L. Knight, 32–35. Madison: University of Wisconsin Press.

Knight, R. L., and K. Gutzwiller, eds. 1995. *Wildlife and recreationists: coexistence through research and management.* Washington, D.C.: Island Press.

Leopold, A. 1933. *Game management.* New York: Charles Scribner's Sons.

Leopold, A. 1938. Report on Huron Mountain Club. Huron Mountain Club, Pine River, Michigan.

Leopold, A. 1939. A biotic view of land. *Journal of Forestry* 37:727–730.

Leopold, A. 1941. Wilderness as a land laboratory. *Living Wilderness* 6(July):3.

Leopold, A. 1949. *A sand county almanac and sketches here and there.* New York: Oxford University Press.

Leopold, A. 1953. *Round river: from the journals of Aldo Leopold.* L. B. Leopold, ed. New York: Oxford University Press.

Leopold, A. 1966. *A sand county almanac with other essays on conservation from Round River.* New York: Oxford University Press.

Lovejoy, T. E. 1980. Foreword. In *Conservation biology: an evolutionary-ecological perspective*, ed. M. E. Soulé and B. A. Wilcox, ix–x. Sunderland, Mass.: Sinauer.

Margolis, J. 1998. The latest 1,000-pound gorilla. *High Country News* 30(8):15.

Meine, C. 1988. *Aldo Leopold: his life and work.* Madison: University of Wisconsin Press.

Noss, R. F. 1992. The Wildlands Project: land conservation strategy. *Wild Earth* (Special Issue):10–25.

Noss, R. F., and A. Y. Cooperrider. 1994. *Saving nature's legacy: protecting and restoring biodiversity.* Washington, D.C.: Defenders of Wildlife and Island Press.

Soulé, M. E., and B. A. Wilcox, eds. 1980a. *Conservation biology: an evolutionary-ecological perspective.* Sunderland, Mass.: Sinauer.

Soulé, M. E., and B. A. Wilcox. 1980b. Conservation biology: its scope and challenge. In *Conservation biology: an evolutionary-ecological perspective*, ed. M. E. Soulé and B. A. Wilcox, 1–7. Sunderland, Mass.: Sinauer.

Stanley, T. R. 1995. Ecosystem management and the arrogance of humanism. *Conservation Biology* 9:255–262.

United Nations Educational, Scientific, and Cultural Organization (UNESCO). 1974. Task force on criteria and guidelines for the choice and establishment of biosphere reserves. *Man and the Biosphere Report No. 22.*

Chapter 8

Professor Leopold, What Is Education For?

The question is, does the educated citizen know he is only a cog in an ecological mechanism? That if he will work with that mechanism his mental wealth can expand indefinitely? But that if he refuses to work with it, it will ultimately grind him to dust? If education does not teach us these things, then what is education for?

Aldo Leopold

WINIFRED B. KESSLER AND ANNIE L. BOOTH

In Aldo Leopold's view, the essential purpose of education is to inform citizens about their place in the ecosystem as a basis for the intelligent and sustainable use of lands and natural resources. This idea winds like a thread throughout the essays of Leopold's *A Sand County Almanac.* Leopold seemed to have a broad educational purpose in writing the essays: to inspire understanding, respect, and ethical reflection to counter society's destructive tendencies toward land and natural resources. Finch (1989, xxi), while noting the bitterness and pain in these essays, concluded that Leopold "remained convinced that most environmental mistakes are due, not to some inherent baseness in human nature but to ignorance. He understood that his own ability to perceive and understand how nature works was the result of a long period of education and self-education." Leopold envisioned a broad and inclusive role for educators in society and for environmental education in the lives of the world's citizens.

For those of us who teach, Leopold's words are a challenge to continually examine our own knowledge and feelings about human-land relationships and to bring deep understanding into our roles as educators. Leopold saw a need for all people to understand their place in the natural world and to embrace land stewardship as a personal ethic. This view challenges us to extend our educational roles well beyond the boundaries of our universities and disciplines. Contained within *A Sand County Almanac* and Leopold's other writings are ideas on how such educational aims might be achieved. Leopold's views on education included: a holistic approach to ecology and natural resource management, recognition of the need for application and personal experience to reinforce learning, connections to the land, integration of humanistic and scientific values, and ethical reflection. Although not the first or the only place where such ideas have been expressed, Leopold's writings are unique in being so widely read and accessible.

We examine the key ideas about education that Leopold conveyed in his writings and actions. We then speculate about how one of the most significant issues of the last half-century might have played out differently if Leopold's ideas had become ingrained in the educational mainstream fifty years ago. Finally, we share our perspectives on the manner in which Leopold's ideas are used in natural resource education today.

Professor Leopold

Although highly respected and liked as a professor at the University of Wisconsin, Leopold was critical of contemporary university education and skeptical of its abilities to slow the deterioration of land and resources. In his "Good Oak" essay, the irony is hard to miss: "Now we cut 1910, when a great university president published a book on conservation, a great sawfly epidemic killed millions of tamaracks, a great drouth burned the pineries, and a great dredge drained Horicon Marsh" (Leopold 1989, 11).

Leopold considered the land to be far more complex than a set of resources or a collection of individual plants and animals. He recognized the importance of ecological interactions and relationships. With respect to education, he was concerned about the fragmentation that existed among the different disciplines: "Education, I fear, is learning to see one thing by going blind to another" (Leopold 1966, 168). His views on the need for integration and holism in education mirrored his ideas about the fundamental nature and workings of ecological systems. Leopold expressed his deep concern about the increasing specialization of professors in this criticism of universities: "There are men charged with the duty of examining the construction of the plants, animals, and soils which are the instruments of the great orchestra. These men are called professors. Each selects one instrument and spends his life taking it apart and describing its strings and sounding boards. This process of dismemberment is called research. The place for dismemberment is called a university" (Leopold 1966, 162).

Leopold viewed a connection to the land as a necessary grounding for developing an understanding of how nature works and how it responds to human interventions. The time and effort invested in an old, abandoned farm on the Wisconsin River would eventually inspire the essays of *A Sand County Almanac.* One of the most frequently cited passages in the *Almanac* addresses the importance of connection: "I have read many definitions of what is a conservationist, and written not a few myself, but I suspect that the best one is written not with a pen, but with an axe. It is a matter of what a man thinks about while chopping, or while deciding what to chop" (Leopold 1989, 68). Likewise, in his essay "Natural History," Leopold (1966) expressed concern that much of what is taught is divorced from real-world application.

Leopold's philosophy on education emphasized the need to include humanistic, as well as scientific, values, an approach that set him apart from both his predecessors and his contemporaries (Finch 1989, xxii). In his essay "The Geese Return," Leopold described how he and his students determined that the lone geese they observed during spring migration were survivors whose mates had likely been killed in the previous hunting season. "Now I am free to grieve with and for the lone honkers. It is not often that cold-potato mathematics thus confirms the sentimental promptings of the bird-lover" (Leopold 1966, 22). Leopold's characterization of university faculty offered a biting critique of the separation of science from the humanities: "A professor may pluck the strings of his own instrument, but never that of another, and if he listens for music he must never admit it to his fellows or students. For all are restrained by an ironbound taboo which decrees that the construction of instruments is the domain of science, while the detection of harmony is the domain of poets" (Leopold 1966, 162).

Leopold's teachings included the need for ethical reflection on natural resource decisions and actions and the flexibility to reconsider what may be "right" or "wrong" in view of those reflections. In "Thinking Like a Mountain," he related an experience with a dying wolf that caused him to reflect on the wisdom of shooting these animals on sight (Leopold 1989). These reflections eventually led him to reformulate his views on predator control.

Ever the student himself, he recognized the need to continually question his own precepts of right and wrong in light of experience and deep contemplation about the workings of the world. In fall of 1938, Leopold began to sign his letters "Professor of Wildlife Management" instead of "Professor of Game Management" (Meine 1988); his reflections about the changing purpose and purview of the profession extended to his own self-identity.

A key message in all of Leopold's writings was the need for humility, respect, and love in people's relationships with nature. Leopold's land ethic relied upon individuals developing an "ecological consciousness" and assuming responsibility for the land. He considered love for the land to be a necessary element of a land ethic: "We can be ethical only in relation to something we can see, feel, understand, love, or otherwise have faith in" (Leopold 1989, 214); "It is inconceivable to me that an ethical relation to land can exist without love, respect, and admiration for land, and a high regard for its value" (Leopold 1989, 223). Leopold was also aware that ecological understanding—including respect and love for the land—came at a high cost: "One of the penalties of an ecological education is that one lives alone in a world of wounds. . . . An ecologist must either harden his shell and make believe that the consequences of science are none of his business, or he must be the doctor who sees the marks of death in a community that believes itself well and does not want to be told otherwise" (Leopold 1966, 197).

How would Leopold feel if he could examine universities today with respect to his critique of fifty years ago? He would probably have mixed emotions. Significant conservation gains have occurred in land-use policies, programs, and practices in North America, such as adoption of the Endangered Species Act, designation of Wilderness Areas, and establishment of the Conservation Reserve Program. However, these gains more closely reflect the activism of individuals and interest groups than a movement championed and nurtured by universities to institutionalize an ecological consciousness. We expect that Leopold would be dismayed to find that, still today, "We may, in fact, be training *idiot savants*—individuals skilled in certain areas (in this case, the technical biological aspects of conservation) but largely inept in other aspects of the field" (Jacobson and McDuff 1998, 263). The focus on individual disciplines and subjects rather than the integration of the whole continues to prevail in natural resource education. If you doubt this, consider the fate of a professor today who is so foolish as to specialize in interdisciplinary research before becoming tenured, or who might apply with an interdisciplinary degree to a wildlife biology, forestry, geography, political science, or economics department.

The way we teach ecology today, especially with the de-emphasis on natural history and field observation, would certainly give Leopold pause. He would find that, although students may spend a few weeks doing fieldwork (usually in good weather), most of their education takes place inside a building. The belief that most learning is from books, lectures, and computer screens, rather than from personal experience in real-world problems, would undoubtedly concern Leopold, who so highly valued a connection with the land.

Leopold would find that the integration of scientific observation with humanistic values, so prominent in his own writings, is still excluded in "mainstream" educational programs in natural resources. Fifty years after publication of *A Sand County Almanac,* we seldom find ethical reflection and other humanistic qualities as learning objectives or instructional outcomes in the natural resource curricula (Kessler 1995). As in Leopold's day, discussions about right and wrong are largely relegated to religious studies and philosophy departments. The enlightened "love" of the land urged by Leopold cannot be mandated as required learning.

What If . . . ? Retrospection on the Spotted Owl

How might things be different today if Leopold's ideas on education had taken hold and flourished since *Almanac* was published? The possibilities for speculation are endless. Let's consider one of the most difficult and controversial issues since that time—the conflict between spotted owl conservation and forest harvesting—and imagine how it may have been influenced by Leopold's views on education. The history, causes, and lessons of this issue, thoroughly documented by Yaffee (1994), provide broad insights into the character of disputes in natural resources today.

Analysis of the spotted owl controversy revealed several reasons for its severity and persistence, including the character of the issue itself. The issue was complex; the stakes were high for the different interests; there were uncertainties in the science, economics, and policy; and little room existed to craft a solution by the time the parties were willing to look for one (Yaffee 1994). Like most controversies in natural resource and environmental sciences, this issue turned out to involve a "multitude of sub-issues, so that seemingly simple choices [became] battles over a variety of other substantive, organizational, and political objectives" (Yaffee 1994, 156). Although an important issue in its own right, the spotted owl dispute was also a symptom of greater, more complex concerns involving old-growth forests, resource-dependent communities, and the credibility of government bureaucrats and agencies.

The stakes in the spotted owl issue were immense. From an economic perspective, the old-growth forests required by the owl were the only source of timber to keep logging and milling operations and resource-dependent communities operating in the region. From an ecological perspective, the old forests represented a scarce, world-class biological resource that would be irreplaceable when cut. The media's portrayal of the issue in black and white—people were either "for jobs and people" or "for the owl"—created extreme polarization and paralysis. Incomplete information and the urgency for decisions created anxiety and uncertainty for all involved. Initially, politicians viewed the "owl problem" as a simple matter that scientists, given enough research, would be able to fix. It turned out to defy solution, however, because "under its surface were clashes of fundamental human values, that is, differences of opinion that are deeply held and unresolvable on objective ground" (Yaffee 1994, 178).

Fragmentation in administrative authority, in land-use policies, in political jurisdictions, and in decision-making processes was a significant barrier to progress in achieving resolution. Attitudes of policymakers ("it's not my problem"), managers ("my hands are tied"), and bureaucrats and legislators ("we're dealing with it" [when we're really not]) exacerbated the difficulties (Yaffee 1994). Along the way, the chief agency involved, the U.S. Forest Service, came to realize that its institutional values, norms of behavior, and culture were ineffectual—as were its organizational structure and administrative processes.

How might things have differed if Leopold's ideas had been infused, through the universities, into an entire generation of natural resource professionals? To begin with, a holistic and integrated perspective might have enabled them to grasp the spotted owl issue—and all that it symbolized—in its entirety. Having been taught that complexity is the norm in ecology and natural resource management, the professionals would have recognized that this was not a simple biological problem requiring a technical fix. Presumably, a full array of social scientists, economists, ecologists, policy specialists, and others would have been enlisted from the start as members of the problem-solving team.

Those who were connected to the problem—including scientists, managers, educators, and affected citizens—would have been engaged early on to find common ground and search for alternative solutions. Thanks to decades of environmental education and outreach efforts by university faculty, natural resource professionals, and others, public understanding of ecology and natural resources would have been strong enough to reject the media's characterization of the issue as black and white (jobs vs. owls). We might even speculate that members of the media might have become ecologically literate enough to cover such issues in an informative and useful manner!

Resource professionals, many of whom filled high administrative and policymaking positions by the 1980s, would have reflected deeply on the issue with respect to their own ecological consciousness and ethical outlook on land, resources, sustainability, and professional responsibility. Those who shirked responsibility or distorted the issue would not have been tolerated by the ethical and committed majority. Having received a natural resource education that included humanistic, as well as scientific, values, these leaders would have been more perceptive and empathetic toward the diverse views that composed the spotted owl dispute. The various interests, in turn, would have placed more trust and confidence in their leaders to handle the issue in a fair, responsible, and intelligent manner.

If Leopold's wish had really come true—and the diverse publics had developed understanding, respect, and love for the land—we might speculate that the spotted owl issue would never have come about. The notion of managing forests so as to maintain their ecological integrity and biodiversity ("every cog and wheel") would have become conventional wisdom by the late 1970s, when the issue began to heat up. Of course, the possibilities for speculation are endless!

Leopold in the Classroom Today

How are the ideas of Leopold used by educators today? We assert that the application of Leopold's ideas, while common in natural resource and environmental curricula, represents a selective process rather than an embodiment of the whole. Educators tend to select those ideas that are compatible with accepted scientific paradigms or with their comfort levels. For decades, wildlife professors have exposed students to the principles and management practices contained in *Game Management* (Leopold 1933) to provide historical context and to illustrate adaptation of practice in the face of new scientific information. Today, a popular use of Leopold's ideas derives from his ecological views on land use: "A thing is right when it tends to preserve the integrity, stability, and beauty of the biotic community. It is wrong when it tends otherwise" (Leopold 1989, 224–225). This statement expresses the fundamental idea (termed "ecosystem management" in today's lexicon) that management should strive to maintain the ecological integrity of forests, rangelands, and other ecosystems.

Of the many ideas that might be explored with students in ecology and natural resources, we suspect that Leopold's ideas on humility, respect, and, most of all, love of the land are too much of a stretch for the comfort level of most professors. Love of the land is a most difficult element to incorporate into resource management plans and practices, and yet it accounts for much of the emotionally charged conflict that surrounds land-use planning and decisions. Although educators should not tell students what to believe or how to feel, they should encourage students to engage in deep, ethical reflection on relationships of people, land, and natural resources (Kessler 1995).

The design of our curriculum at the University of Northern British Columbia embodies several of Leopold's ideas on education. Beginning with their first-year courses, students in the Bachelor of Science program in natural resource management receive a thoroughly integrated perspective that includes the ecological, economic, and sociocultural dimensions of natural resources. A fourth-year course in environmental and professional ethics is required of all students in the program, including majors in forestry, wildlife, fisheries, and resource recreation. This course, NREM 411, has acquired a reputation as the "killer course" in the Natural Resource Management curriculum because it requires students to demonstrate ways of learning and expressing themselves that are quite different from what they have proven adept at in their other courses. When asked to analyze the ethical implications of an issue, many students prefer instead to provide a factual account of the issue or to cite what other people (the "ethics specialists") have to say.

They soon learn that this approach will not suffice in NREM 411, in which the objectives are for every student to reflect deeply on the ethical implications of issues pertaining to natural resources, to find and defend their personal position within the spectrum of ethical debate, and to develop a personal creed that will shape their conduct in professional life.

We often find that students in NREM 411, even those who earn excellent grades in their science, management, and policy classes, have a hard time in this course. If Leopold were here, he might attribute these difficulties to a mechanical, detached approach to education that inadequately encourages personal reflection and deep contemplation of the wider implications of human actions. We would have to agree with him.

References

Finch, R. 1989. The delights and dilemmas of *A Sand County Almanac.* In sA. Leopold, *Special commemorative edition, a sand county almanac and sketches here and there*, xv–xxviii. New York: Oxford University Press.

Jacobson, S. K., and M. D. McDuff. 1998. Training idiot savants: the lack of human dimensions in conservation biology. *Conservation Biology* 12:263–267.

Kessler, W. B. 1995. Wanted: a new generation of environmental problem-solvers. Wildlife Society Bulletin 23:594–599.

Leopold, A. 1933. *Game management.* New York: Charles Scribner's Sons.

Leopold, A. 1966. *A sand county almanac with other essays on conservation from Round River.* New York: Oxford University Press.

Leopold, A. 1989. *Special commemorative edition, a sand county almanac and sketches here and there.* New York: Oxford University Press.

Meine, C. 1988. *Aldo Leopold: his life and work.* Madison: University of Wisconsin Press.

Yaffee, S. L. 1994. *The wisdom of the spotted owl: policy lessons for a new century.* Washington, D.C.: Island Press.

Chapter 9

Aldo Leopold and the Value of Nature

STEPHEN R. KELLERT

When questions arise about the value of healthy natural systems to human well-being, they are typically framed in the context of economic and sometimes ecological benefits. This is not surprising given our market-driven and scientifically oriented society. There is a tendency for issues of environmental concern to be decided on the basis of empirical evidence and other quantifiable, presumably objective, measurements of value. This prevailing emphasis on quantifiable and economic benefits tends to obscure those difficult-to-quantify, personal relationships to the natural world that have traditionally shaped the human experience. Moreover, the value of nature is seen as independent of human culture and history.

By contrast, when we examine environmental value systems in their cultural context, we find a great diversity of meanings and benefits attributed to nature. This "relativist" viewpoint particularly emphasizes what seem like distinctive aesthetic, ethical, and spiritual perspectives in assigning value to nature among particular individuals, cultures, and historical circumstances. This relativist position typically endorses three assumptions about nature and humanity. First, it sees people's environmental perceptions as extraordinarily variable and pliable, the product of social conventions, power structures, and distinctive historical circumstances or "discourses." Second, it views humans as possessing an unusual, perhaps unique, capacity to transcend the dictates of our species' heredity and biology. Finally, the relativist position suggests that the values placed on nature by different peoples at particular places and times are neither "good" nor "bad" in and of themselves. Instead, such "value constructions" are seen as the reflection of certain power and economic relationships prevailing at the time.

The work of Aldo Leopold helped me to reconcile the seeming contradiction between the universal and relative expression of human values in regard to nature. In the process, I gained a better understanding of how environmental values relate to human evolution and biology, as well as to an ethic for conserving natural systems. Leopold realized that both universal and relative perspectives contain important elements of truth and, more significantly, that these truths are not mutually exclusive. Leopold recognized that considerable variability occurs in people's perceptions of nature, but within a framework originating in our biological and genetic dependence on the natural world. In advancing this understanding, he went beyond a shallow, economic utilitarianism, reaching much deeper and more profoundly to an understanding of how even aesthetic, emotional, and moral understandings of the natural world fundamentally reflect our biological dependence on natural systems and processes.

We can understand Leopold's perspective by examining his views on the human aesthetic response to nature. An aesthetic value is particularly relevant here because many view perceptions of beauty or physical attraction to the natural world as capricious, merely reflecting prevailing tastes, fashions, or cultural bias. As a consequence, aesthetic views of nature are often regarded as only marginally relevant to issues of human well-being and irrelevant to considerations of evolutionary fitness, despite the apparent ubiquity of an aesthetic attraction to nature among seemingly all peoples and cultures. Leopold, instead, affirmed the importance of the aesthetic response to nature as essential to human well-being, and he intimated its biological, evolutionary, and developmental significance. He suggested, for example:

> Our ability to perceive quality in nature begins, as in art, with the pretty. It expands through successive stages of the beautiful to values as yet uncaptured by language. The quality of cranes lies, I think, in this higher gamut, as yet beyond the reach of words. . . . When we hear his call, we

> hear no mere bird. We hear the trumpet in the orchestra of evolution. He is the symbol of our untamable past, of that incredible sweep of millennia which underlies and conditions the daily affairs of birds and men. (Leopold 1966, 96)

He also stated:

> The physics of beauty is one department of natural science still in the Dark Ages. . . . Everybody knows . . . that the autumn landscape in the north woods is the land, plus a red maple, plus a ruffed grouse. In terms of conventional physics, the grouse represents only a millionth of either the mass or the energy of an acre. Yet subtract the grouse and the whole thing is dead. An enormous amount of some kind of motive power has been lost. (Leopold 1966, 137)

More than half a century later, we continue to struggle to capture in words and through science the sense of the deep biological importance of our intuitive response to nature, including our sense of beauty in the natural world. In remarks like those above, Leopold reached beyond a simple and narrow, utilitarian calculation of nature's worth to humanity, invoking considerations of not only aesthetics, but also kinship and morality. He did this again in commenting on the extinction of the passenger pigeon:

> We grieve because no living man will see again the onrushing phalanx of victorious birds sweeping a path for spring across the March skies, chasing the defeated winter from all the woods and prairies. . . . Our grandfathers were less well-housed, well-fed, well-clothed than we are. The strivings by which they bettered their lot are also those which deprived us of pigeons. Perhaps we now grieve because we are not sure . . . that we have gained by the exchange. The gadgets of industry bring us more comforts than the pigeons did, but do they add to the glory of the spring? It is a century now since Darwin gave us the first glimpse of the origin of species. We know now what was unknown to all the preceding caravan of generations: that men are only fellow-voyagers with other creatures in the odyssey of evolution. This new knowledge should have given us, by now, a sense of kinship with fellow-creatures . . . a sense of wonder over the magnitude and duration of the biotic enterprise. (Leopold 1966, 109)

While not minimizing the importance of economic benefits derived from natural systems, Leopold extended this understanding to include aesthetic, emotional, and spiritual values as well, each possessing an adaptive and biological significance to human physical and mental well-being. Thus, Leopold placed the human perception of value in nature squarely and fundamentally in an ecological and evolutionary context. In one of his most frequently cited statements in this regard, he explicitly and boldly articulates a land ethic that links environmental values to human biology and includes both economic and noneconomic dependencies:

> An ethic to supplement and guide the economic relation to land presupposes the existence of some mental image of land as a biotic mechanism. We can be ethical only in relation to something we can see, feel, understand, love, or otherwise have faith in. . . . Ethics . . . is actually a process in ecological evolution. . . . An ethic may be regarded as a mode of guidance for meeting ecological situations so new or intricate, or involving such deferred reactions, that the path of social expediency is not discernible to the average individual. . . . Ethics are possibly a kind of community instinct in the making. . . . The 'key-log' which must be moved to release the evolutionary process for an ethic is simply this: quit thinking about decent land-use as solely an economic problem. Examine each question in terms of what is ethically and esthetically right, as well as. . . . economically expedient. A thing is right when it tends to preserve the integrity, stability, and beauty of the biotic community. It is wrong when it tends otherwise. (Leopold 1966, 230, 218–219, 240)

These views were instrumental in my development of a typology of basic values people hold in regard to nature and eventually my connection of these values to an understanding of human evolutionary development and well-being (Kellert 1996, 1997). In constructing this typology, I wanted to go beyond physical and material benefits people derive from natural systems to include aesthetic, emotional, moral, and other noncommodity benefits as well. Leopold's views helped me to see the universal, evolutionary, and biological origins of these values humans perceive in nature, despite the wide diversity of their expression among individuals, groups, and cultures.

Biophilia and Environmental Values

I was also influenced in this regard by the work of Edward O. Wilson, particularly concerning the notion of "biophilia," which suggests that people possess a genetic and broadly based inclination to value the natural world (Wilson 1984; Kellert and Wilson 1993). Like Leopold, Wilson recognized that humans have inherent psychological affinities to the natural world, including aesthetic appreciation, emotional attachment, and spirituality, and all of these affinities have evolutionary and developmental significance. For example, in arguing for the protection of biological diversity, Wilson remarked in ways reminiscent of Leopold:

> Decisions concerning . . . biodiversity will turn on our values and ways of moral reasoning. A sound ethic . . . will . . . take into account the immediate practical uses of species, but it must reach further and incorporate the very meaning of human existence. . . . A robust, richly textured, anthropocentric ethic can . . . be made based on the hereditary needs of our own species, for the diversity of life based on aesthetic, emotional, and spiritual grounds. (Wilson 1993, 37)

In my work, I have linked the concept of biophilia to nine biologically based environmental values, including the human need for material sustenance, the quest for empirical understanding, the desire for emotional connection, the capacity for critical thinking, the urge for exploration and discovery, the search for spiritual meaning, and more. These nine values are briefly identified in the table below (Table 9.1). Each reflects an inherent human affinity for the natural world of relevance to human evolutionary fitness and physical and mental well-being. Despite the universal expression of these values, they are, nonetheless, seen as "weak" biological tendencies, greatly influenced and shaped by culture, learning, and experience. Such variability has its positive dimensions, reflecting the creativity, free will, and diversity of human perceptions of nature across individuals, groups, and cultures. But this variation is also bounded, dictated by our biology. If these values fail to manifest themselves, if they atrophy or become abnormally exaggerated, individuals and even whole societies will suffer.

These nine values express the complexity of the human reliance on the

TABLE 9.1. A typology of values associated with the biophilia hypothesis.

Aesthetic: physical attraction and appeal of nature.
Dominionistic: mastery and control of nature.
Humanistic: emotional bonding with nature.
Naturalistic: exploration and discovery of nature.
Moralistic: moral and spiritual relation to nature.
Negativistic: fear of and aversion to nature.
Scientific: knowledge and understanding of nature.
Symbolic: nature as a source of communication and imagination.
Utilitarian: nature as a source of material and physical rewards.

natural world for physical, emotional, intellectual, and spiritual sustenance and security. Conversely, as both Leopold and Wilson acknowledged, degrading or diminishing this experience and dependence on nature can and will impoverish the human capacity for achieving material security, personal identity, social stability, and moral worth. A broad, instrumental ethic for sustaining the health, integrity, and beauty of the biota, thus, depends on the functional and adaptive expression of all values of biophilia. Even our fears of and aversions to elements of nature can serve to encourage a recognition of the power of creation. By contrast, species and habitats utterly subdued rarely evoke our admiration, humility, and respect. Ethical regard for the natural world derives not just from feelings of compassion or the need to ensure our material well-being, but from a broader awareness of how nature shapes the human body, mind, and spirit. Leopold implied this broader view of conservation when he remarked:

> Conservation is a state of harmony between men and land. . . . You cannot cherish [the] right hand and chop off [the] left. That is to say, you cannot love game and hate predators; you cannot conserve the waters and waste the ranges; you cannot build the forest and mine the farm. . . . If the land mechanism as a whole is good, then every part is good, whether we understand it or not. If the biota, in the course of aeons, has built something we like but do not understand, then who but a fool would discard seemingly useless parts? (Leopold 1966, 176–177)

Human Well-Being in Nature

This brief essay has articulated some of the ways in which Aldo Leopold's work influenced my scholarly development. In concluding, I want to emphasize two other related aspects of Leopold's influence on my thinking. First, Leopold helped me to appreciate that we must include and elucidate the role of humans in natural systems if we are to understand and sustain these systems. At an early stage in my career, this realization caused me to pay attention to the often heard, but rarely examined, refrain, "Wildlife management is really people management." I particularly took to heart Leopold's assertion that "The [wildlife] problem is not how we shall handle [species]. . . . The real problem is . . . human management. Wildlife management is comparatively easy; human management difficult" (Flader 1974, 88). I concentrated my efforts on exploring human attitudes and behaviors toward wild creatures and their habitats and how these relationships were distributed across diverse demographic and interest groups in the United States and elsewhere.

It was not long before I began to discover consistent patterns of human relationships to nature among all those studied. This encouraged me to pursue Leopold's "deeper vision" of humans as an integral part of nature, possessing, like any organism, the potential to exercise positive, as well as negative, effects. Leopold helped me to see human values as a fundamental shaping influence on the structure and functioning of natural systems, rather than as some kind of exogenous, alien, or added layer to the natural world. I began to view environmental values from a fundamentally ecological and evolutionary perspective.

Leopold's work further prompted me to focus on the role of individual experience and personal self-interest in understanding how natural systems shape humanity and vice versa. This, in turn, led me to an awareness of how people depend on the health and integrity of natural systems for physical, mental, and spiritual well-being. As a consequence, Leopold's work legitimized and reinforced in me an inclination to look at human-nature relations from the perspective of a "naturalist." Like Leopold, I came to believe that understanding the fundamental interconnection and interdependence of humans and nature required intimate and ongoing affiliation with the wonder, beauty, and complexity of the natural world. This meant developing the willingness to become a keen and enthusiastic observer and, just as important, to pursue the personal satisfaction through deep, participatory involvement in nature, whether as a researcher, birder, hunter, hiker, or conservationist.

I became committed to the view that we must understand the role of nature in "everyday life." If the experience of nature exercises a fundamental shaping influence on human physical and mental well-being, then this experience must be understood in the context of normal, everyday relationships. As a consequence, I developed a growing interest in examining the developmental effects of nature during childhood and in better understanding the connections between the human-built and natural environments. These interests have recently led me to consider the design of human structures and communities that acknowledge and seek to reinforce the many affirmative values people derive from contact with healthy natural systems.

This emphasis on the personal experience of nature also led me to the conviction, as suggested earlier, that an ethic for sustaining the health and integrity of the natural world ultimately depends on a profound realization of self-interest. Like Leopold, I came to realize how much nature shapes our identities and offers us opportunities to achieve lives of meaning, beauty, and grace. The natural world remains the bedrock of our material security, intellectual capacity, emotional bonding, and spiritual connection. Conversely, we cannot achieve lives of physical and moral worth built upon the destruction of nature as we know it.

Leopold suggested "the outstanding scientific discovery of the twentieth century is not television [or the computer] . . . but rather the complexity of the land organism" (Leopold 1966, 176–177). He recognized that, whether as individuals, as communities, or as a species, we remain products of our relational dependencies. The notion of a single, autonomous being is largely an illusion. Our unyielding and complicated ties to creation underscore the necessity of a broad, anthropocentric environmental ethic. We are spiritually nourished and morally guided by celebrating our commonality with nature. Degrading this relationship engenders more than material harm; it inevitably leads to an impoverished spirit, a loss of moral bearings, and an eroded identity. In his musings on wolves and mountains, Leopold intimated this human need to achieve unity and integrity with the natural world:

My own conviction on this . . . dates from the day I saw a wolf die. . . . In those days we never heard of passing up a chance to kill a wolf. . . . We reached the old wolf in time to watch a fierce green fire dying in her eyes. I realized then, and have known ever since, that there was something new to me in those eyes—something known only to her and to the mountain. I was young then, and full of trigger-itch; I thought that because fewer wolves meant more deer, that no wolves would mean hunters' paradise. But after seeing the green fire die, I sensed that neither the wolf nor the mountain agreed with such a view. Since then I have lived to see state after state extirpate its wolves. I have watched the face of many a newly wolfless mountain. . . . I have seen every edible bush and seedling browsed, first to anaemic desuetude, and then to death. . . . In the end the starved bones of the hoped-for deer herd, dead of its own too-much, bleach with the bones of the dead sage, or molder under the high-lined junipers. I now suspect that just as a deer herd lives in mortal fear of its wolves, so does a mountain live in mortal fear of its deer. . . . Perhaps this is the hidden meaning in the howl of the wolf, long known among mountains, but seldom perceived among men. (Leopold 1966, 129–133)

References

Flader, S. 1974. *Thinking like a mountain: Aldo Leopold and the evolution of an ecological attitude toward deer, wolves, and forests.* Columbia: University of Missouri Press.

Kellert, S. 1996. *The value of life: biological diversity and human society.* Washington, D.C.: Island Press.

Kellert, S. 1997. *Kinship to mastery: biophilia in human evolution and development.* Washington, D.C.: Island Press.

Kellert, S., and E. O. Wilson, eds. 1993. *The biophilia hypothesis.* Washington, D.C.: Island Press.

Leopold, A. 1966. *A sand county almanac, with other essays on conservation from Round River.* New York: Oxford University Press.

Wilson, E. O. 1984. *Biophilia: the human bond with other species.* Cambridge, Mass.: Harvard University Press.

Wilson, E. O. 1993. Biophilia and the conservation ethic. In *The biophilia hypothesis,* ed. S. Kellert and E. O., Wilson, 31–41. Washington, D.C.: Island Press.

A Land Ethic in Practice

Chapter 10

Leopold's Land Ethic

A Vision for Today

JAMIE RAPPAPORT CLARK

Restoring a Missing Piece

26 January 1998. The crisp winter day greeted us with a fresh blanket of snow covering the mountain landscape. Ponderosa pines reached for the clear, cerulean-painted sky, while the vistas atop the summit offered million-dollar views. From our vantage point, we could see forever. This winter morning demanded a celebration, of what we were planning to do that day—restore a missing piece of our nation's natural heritage.

I, along with Secretary of the Interior Bruce Babbitt, was at the Apache National Forest in eastern Arizona to release the endangered Mexican gray wolf back into the wilderness of the desert Southwest. After an absence of nearly fifty years, the wolf would be returning home to roam once again amongst this rugged landscape. For the past nine years, I had worked to make this day a reality; it became such a passion that I named my golden retriever Bailey as a reminder of my goal to put *Canis lupus baileyi* back into its rightful place in the wild. Just like what we had done a few years earlier to restore the gray wolf in Yellowstone National Park and Idaho, the time had come to bring back the howl of the wolf to the Desert Southwest.

On that beautiful day in Arizona's Blue Range Mountains, I felt proud of what we had worked so long to accomplish, and I also sensed the significance and symbolism of what was about to happen. Peering into the crates holding the wolves, I remembered the words Aldo Leopold wrote in his essay "Thinking Like a Mountain": "Only the mountain has lived long enough to listen objectively to the howl of a wolf" (Leopold 1966, 129). Many of us on the mountain that day had wished for a long time that Americans could once more hear the voice of this wilderness symbol, and now we knew that the mountains would agree with what we were about to do. The desert's voice would be returning. All it was going to take now was the opening of the crates holding these symbols of wildness.

As Secretary Babbitt raised the sliding doors on the crates, the wolves darted from the confines of their temporary imprisonment. After they entered the acclimation enclosure, I could feel the mountains breathe a sigh of relief. Although the wolves were still under lock and key, later they would be permitted to leave the enclosure for the wilds of the Blue Mountains and another piece of nature's puzzle would be back in its proper place amidst this rugged landscape.

The reintroduction of wolves to Arizona is only one example of the many projects around the country attempting to reintroduce native species, to restore declining or degraded ecosystems, and to reinstill natural processes into our nation's wildlands. For me, the release of the Mexican gray wolf in the Arizona wilderness became an example of the hard-earned successes of the Endangered Species Act. The act has saved not only well-known species such as the Mexican gray wolf from extinction, but also significant numbers of lesser-known species—from plants to amphibians, reptiles, birds, and mammals—whose contributions to the overall ecological health of the land remain unknown. One only has to read Leopold's "Round River" essay to understand the significance of even the most minute species:

> The outstanding scientific discovery of the twentieth century is not television, or radio, but rather the complexity of the land organism. Only those who know the most about it can appreciate how little we know about it. The last word in ignorance is the man who says of an animal or plant, "What good is it?" If the land mechanism as a whole is good, then every part is good, whether we understand it or not. If the biota, in the course of eons, has built something we like but do not understand, then who but a fool would disregard seemingly useless parts? To keep every

> cog and wheel is the first precaution to intelligent tinkering. (Leopold 1966, 177)

To me, as a resource professional and a passionate lover of nature, this simple but eloquent passage lays it on the line. It reminds me why we work hard to recover our endangered species, to restore our damaged waterways, and to conserve our public lands. Leopold's essays also emphasized, and rightly so, why we strive not only to restore wildlife populations and their habitats, but also to provide opportunities for the American public and people around the world to revel in the splendor of our nation's natural heritage.

Leopold was clearly ahead of his time. For years, his ideas were not widely incorporated into natural resource management. Even today, there are pockets of resistance to managing wildlife resources as entities of an ecosystem, preferring instead to continually focus on managing one species at a time. Fortunately, the tide is turning and, for many resource professionals that have a better grasp of basic ecological concepts, they know that every component of the system, large or small, plays a role in preserving and maintaining the integrity of an ecosystem.

For the contemporary resource professional, to take an ecosystem approach to resource management requires not only dedication, skills, and knowledge, but also a heartfelt passion for all things natural, wild, and free. It also requires a new way of implementing wildlife conservation. It requires more public involvement and education. It requires an understanding of social sciences and the importance of communication. The days of the field biologist happily trudging through the marsh in hip waders with nary a worry, except for the numbers and types of wildlife counted, are over. No longer can we do business as usual. The stakes are too high, and the challenges are too daunting.

As much as we can celebrate our successes in conserving and restoring our nation's wildlife heritage, the truth is that things still aren't right on the landscape. For some in the profession, it remains a struggle to embrace Leopold's land ethic, to break out of the paradigm of single-species management. Nowhere is this more evident than in the campaign to conserve and restore our nation's endangered species. For the U.S. Fish and Wildlife Service (USFWS), this has meant a change in both our approach to conserving fish, wildlife, and plants, and in our daily operations as an agency.

From a Single-Species to an Ecosystem Approach

Throughout much of its history, the USFWS, as well as other federal land management agencies and state game and fish departments, focused on the management of game species. The training wildlife biologists received in college fostered this single-species approach. Our universities offered wildlife courses specific to management of upland game, forest game, and waterfowl. Not much attention was paid to nongame wildlife, endangered species, and ecological processes. A word or two might be expressed at the end of the course about how what we do for game benefits the songbirds, but that was the extent of it. Management actions for the benefit of the deer, turkey, or squirrel superseded any concern some of us had for degrading or destroying the habitats of those "nonconsumptive" species. Once college graduates entered the career field, they were tasked with developing deer management plans, turkey management plans, and squirrel management plans. The three plans never connected—they remained separate entities. No attention was given to integrating the approaches for each species, to avoiding conflicts and potential damaging results in the competing management schemes carried out, and to considering the effects the management actions taken to enhance these species would have on other species. In college and in the profession, the single-species approach prevailed.

For the USFWS, the primary focus for much of its storied history was waterfowl management. Although a shift away from single-species management to protecting all the "cogs and wheels" of an ecosystem is occurring in many resource agencies, including the USFWS, the traditional attitude still prevails in some corners. The new way of doing business, of looking at the whole landscape and making decisions based on the concepts of landscape ecology, ecosystem science, and conservation biology, remains difficult for many to accept, implement, or even understand.

Today, in a new century, we must learn from past mistakes and not repeat them. We should accept and refine traditional tools that still have merit. Most important, we have to become more innovative and creative in the tools and approaches we use to reverse the decline of wildlife communities and to restore degraded ecosystems. We must, in our management decisions, be inclusive of all living things. The traditional, single-species approach to wildlife management that was instrumental in recovering many of our game species has now evolved into an approach that considers the entire landscape. We must acknowledge the "flux of nature" instead of the traditional "balance of nature." We must incorporate into our management strategies the use of natural processes to recover and restore degraded landscapes.

No one said it would be easy; the new methods of resource conservation are just as demanding and challenging as the techniques our predecessors used in the early days to restore our nation's depleted game populations. Just as it was for them, resource conservation remains a daunting task for us. Instead of just creating and enhancing habitats for wildlife, we must now try to restore landscapes and reinstate vital ecological processes, whether it be fire or predator-prey dynamics.

It is easy for many of us in the profession today to criticize the approaches used by our predecessors to restore our nation's wildlife heritage. That's unfortunate and unfair. What was accomplished by these pioneers was extraordinary and commendable. The issues facing our wildlife heritage today hardly existed sixty years ago. During those early years, many of today's problems, where they did exist, could not be recognized or were not considered problems at all. Little thought was given in the early days to urbanization, outdoor recreation, or wildlife conservation from a landscape perspective. The serious environmental damage from contaminants or the intrusion of invasive, alien species was virtually nonexistent. The accomplishments of early resource professionals are well known; the dramatic population increases in white-tailed deer, turkey, and elk are a few of the great successes. However, in many situations, the approaches used to restore game populations became detrimental to the other "cogs and wheels" of the landscape: developing edge habitat at the expense of protecting and, more important, restoring extensive tracts of intact forest habitat; impounding tidal wetlands for waterfowl at the expense of destroying vital spawning grounds for marine life; introducing exotic species such as the multiflora rose and the salt cedar to enhance game habitat; and releasing nonnative sport fish that would outcompete native species in our rivers and lakes.

During those early days, it was much easier to deal with one species at a time. For some time, the single-species approach worked, but society's impacts and demands on our natural resources became more complicated. Therein lies the challenge for today's wildlife professionals: how to make the transition from creating and enhancing habitats to restoring and protecting landscapes that are the "last of the best and the best of the rest." How do we take the approaches and techniques used successfully in restoring game populations and integrate them into new methods to protect and restore a whole array of plants, animals, and their habitats?

In many of the USFWS programs, most notably the endangered species program, we have learned the critical importance of collaboration, sharing knowledge and stewardship responsibilities within and among USFWS programs. Amazing success has been achieved with endangered species conservation efforts. The extinction of more than twelve hundred species from the earth has been averted. Populations of more than 44% of the species listed under the Endangered Species Act are either stable or increasing. Considering the rapid expansion of the population and economy of the United States in the last 28 years, accompanied by pressures for development that have crowded out native species at an unprecedented rate, this is a record to be proud of.

Although extinction has been averted for some species, the challenge is much greater heading into the twenty-first century. Currently, more than two hundred candidate species await the initiation of the formal listing process. Essential habitats are still being fragmented and degraded, and the human population continues to grow. By the middle of the next century, the population of the United States will have grown by 125 million people; that's fifteen more New York City's. Imagine what our country will look like when we have to house, clothe, feed, and find employment for another 125 million people. Imagine the impact this will have on our forests, rivers, wetlands and the fish, wildlife, and plants that depend on these ecosystems. To meet this challenge, we all must recommit ourselves to Leopold's vision of looking at the land and its constituent parts. We must do this as a profession, but we must ensure that the American public understands and supports a land ethic as well.

Embracing Leopold's Land Ethic

Within the natural resource conservation profession, we have sought to capture Leopold's vision with terms such as ecosystem management, landscape conservation, and "looking at the big picture." Whatever term we use, we are challenged to broaden our individual or organizational frames of reference—to think globally and act locally.

Developing strategies to protect our wildlife resources begins with the basic premises of Leopold's land ethic. Although some adaptation may be required, the land ethic remains as viable and important today as it was fifty years ago. Conservation of our precious natural resources presents us with these challenges:

1. Recognize the dynamic nature of an ecosystem. When implementing management actions at a landscape scale, we must incorporate considerations for ecosystem resiliency and develop strategies that can accommodate unexpected events and natural disturbance regimes.
2. Constantly monitor natural resources and related management actions so that adjustments can be made. This approach involves continual experimentation with management strategies and approaches, with the understanding that decisions may at times have to be made with less-than-complete information. Management approaches and techniques must remain adaptable to change. Adaptive management is essential in cases where the biological information is not complete enough to make informed decisions. As information becomes available through research and experimentation, decisions will become more educated and relevant.
3. Strive to maintain existing native plant and animal populations and restore those that have suffered drastic declines due to human alteration. Let's finally accept the fact that we induce failure when we try to control the natural variation of a system. For biodiversity conservation, we must adapt the management practices to the system, not the other way around. To the extent practical, we must allow natural processes to operate unimpeded. Instead of creating or enhancing artificial habitats for a few high-interest species, we need to focus our actions on restoring and protecting ecosystems and their native communities.
4. Set clear goals and objectives, including targets that can be measured to monitor ecosystem conditions. We must know what we have before we do something with what we've got. Once this is accomplished, we will be better equipped to make decisions on how best to proceed in restoring and protecting our wildlife resources.
5. Incorporate aesthetic concerns and amenity values into our management approaches. Both elements are important to preserve the natural integrity and appearance of the landscape. When the public can understand the connection between the value of maintaining the integrity of a landscape and its role in protecting the welfare of humans, we will add to our number of advocates for the land.

6. Collaborate with diverse publics as informed, active participants in the process. We must develop approaches that meet the needs and interests of relevant stakeholders within our culturally diverse society. Generic public outreach and environmental education programs are no longer effective in today's world. The challenge lies in our ability to reach out to all groups, especially those without a traditional respect for or understanding of the natural world, and work with them to understand and embrace the concepts of good land stewardship.
7. Develop partnerships. We must recognize that we cannot do the job alone. We must depend on collaborative efforts involving all stakeholders to ensure long-term conservation. Successful partnerships are essential, but challenging. Partnerships require adaptability, risk taking, innovations, a shared vision, active participation, and commitment by all parties involved.

The USFWS in the Twenty-first Century

With the new millennium upon us, the USFWS is, indeed, integrating these essential components of Leopold's land ethic into daily operations. Continuing to implement an ecosystem approach to fish, wildlife, and plant conservation within the agency, while respecting the strengths of a traditional program orientation, is a constant challenge. We need a vibrant national wildlife refuge system, a progressive endangered species program, and solid programs for fisheries, migratory birds, law enforcement, and habitat conservation. At the same time, we need to ensure collaboration among our biologists, and with other disciplines within our agency, to address and resolve conservation challenges across diverse ecosystems. There is no question we have recognized that the ecosystem approach to conservation is our future. Landscape-level conservation through cross-program coordination within the USFWS and in partnership with other organizations and individuals is the job of the USFWS; it is the "normal work" of all our employees, a mission to which our individual and collective efforts must contribute.

To help this process along, the USFWS is focusing on building partnerships with other federal agencies, states, local governments, corporations, foreign countries, conservation organizations, tribes, sportsmen's groups, and private landowners. Over the years, many of our most effective conservation efforts have been through partnerships. Whether in habitat conservation plans, refuge comprehensive conservation plans, safe harbor agreements, or aquatic restoration initiatives for river basins, we know that we cannot do the job alone. We must stay committed to working with others to accomplish the increasingly complex task of resource conservation.

In the twenty-first century, the great victories for conservation may be achieved in cooperation with people who know little about resource conservation or have a far different perspective on the management and use of natural resources. To work successfully with these people, we must see what they see; we must be able to convey to them what we see; and, together, we must find common ground.

Essential to our success as a conservation agency is our ability to communicate. Good biology alone will not get the job done. If we cannot communicate clearly and convincingly, we will not be able to achieve our mission. Aldo Leopold and Rachel Carson demonstrated the power of words to mobilize people for conservation, and we have seen many cases where good communication and outreach have made it possible for us to move forward, even in highly controversial cases such as the reintroduction of wolves to Arizona and Yellowstone National Park and California condors to the Grand Canyon.

As in other federal, state, and local conservation agencies, the USFWS is a family of dedicated resource professionals working to conserve our wildlife heritage. Our commitment and passion is not for money, prestige, or visibility—it is a commitment to conservation.

For us, the sight of a soaring condor, a flock of geese in flight, or something as simple as a black-capped chickadee at a backyard feeder stirs our souls and lifts our spirits. For us, wildlife conservation is a sacred responsibility; we have been given stewardship over something precious and irreplaceable.

As those of us in the conservation profession deal with the difficult issues ahead, it is my hope that we will always keep these things in mind. We must have a sure knowledge that some things are priceless; as Leopold observed, although there are many things we can live without, wild creatures and wild places are not among them. Over fifty years ago, Leopold captured this vision in *A Sand County Almanac*. Now the perpetuation of that vision is up to us!

References

Leopold, A. 1966. *A sand county almanac with other essays on conservation from Round River*. New York: Oxford University Press.

Chapter 11

Aldo Leopold

Conservationist and Hunter

After lunch and a nap we decided to lay in some meat, so I killed a greenhead and Carl got 13 quail. We found that in walking through the wild hemp one can gather quite a lot of beans by simply holding the hands cupped and letting the beans rain in from overhead. Killed a couple of coots for trap bait. Killed a duckhawk chasing a heron.

Aldo Leopold

L. DAVID MECH

A Lifelong Passion

This little-quoted entry from Aldo Leopold's *Round River* journals (Leopold 1953, 24–25) may shock the average new-millennium conservationist. By today's standards, the passage is not exactly environmentally correct, and not just because Leopold referred to the majestic peregrine falcon as a "duckhawk."

Of course, all but the extreme purist will overlook the acts described here and committed in 1922 that today would be considered major transgressions. Still, this passage yields significant insight into Aldo Leopold, the person, at age thirty-five. The fact is, Leopold was a hunter "by inclination, upbringing and desire" (McCabe 1987, 124). That is also clear from Leopold's most quoted writing:

> We reached the old wolf in time to watch a fierce green fire dying in her eyes. I realized then, and have known ever since, that there was something new to me in those eyes—something known only to her and to the mountain. I was young then, and full of trigger-itch; I thought that because fewer wolves meant more deer, that no wolves would mean hunters'

paradise. But after seeing the green fire die, I sensed that neither the wolf nor the mountain agreed with such a view. (Leopold 1966, 130)

However, Leopold's much-acclaimed conversion, as conveyed in that passage, both forgives and obscures the fact that he and his companion were the ones who shot the mother wolf and wounded at least one of her pups: "In those days we had never heard of passing up a chance to kill a wolf" (Leopold 1966, 130).

It was like city kids in the 1960s shooting rats at a garbage dump, or today's Florida housewives snap-trapping thousands of mice in their modern homes after a hurricane. Although some present-day conservationists would object even to the recent storm-induced rodent slaughter, most people would consider it acceptable. So, too, in the 1920s, it was with killing wolves, or for that matter, duckhawks, quail, coots, or waterfowl.

But Aldo Leopold did not stop hunting in the 1920s. Even after he wrote the "green eyes" passage in 1944, he continued to hunt (McCabe 1987), although not necessarily wolves. (He did recommend a bounty to reduce wolves in Wisconsin in 1945 [Flader 1974].) His lifetime love of hunting does not surprise wildlife managers, most of whom enjoy the sport themselves. However, Leopold's poetic prose reached far beyond wildlife managers and even today continues to inspire mainstream audiences, calling them to the environmental cause. His *Sand County Almanac* has sold over a million copies.

I dare speculate that a high percentage of Leopold's readers would be dismayed that the same person who wrote, "only the mountain has lived long enough to listen objectively to the howl of a wolf" (Leopold 1966, 129), also penned, "killed a couple of coots for trap bait. Killed a duckhawk chasing a heron." My point is that, like most of his predecessors and contemporaries, but unlike many of today's environmentalists, Aldo Leopold was a conservationist precisely because he was a hunter, trapper, and fisherman.

The loud echo of this integral relationship between an exploitive and protective attitude toward the natural world is heard regularly in the testimonies and writings of sportsmen's groups, game and fish departments, and old-line conservation organizations: "sportsmen are the true conservationists." I am only repeating it here and relating it to Aldo Leopold because I think this voice is beginning to be drowned out and overwhelmed by a growing chorus of urbanized environmentalists, those whom Leopold characterized as never having felt the soil between their toes.

My guess is that this new army of environmentalists, although certainly welcome to the cause, will never produce a person who inspires quite the passion of an Aldo Leopold. I make that guess because, as a hunter myself, I dare think I understand Leopold's passion.

The Hunting Experience

Let me explain the radical difference between the current popular outdoor activities like hiking, camping, canoeing, skiing, and wildlife photography and the consumptive or exploitive practices like hunting, trapping, and fishing. Having indulged in all the activities of both groups, the best way I can describe that difference is to say that the latter group involves an especially high degree of anticipation and excitement, extreme focus, and an intimate interaction with the natural world. That combination produces such a euphoria that a former student of mine likened it to a drug-induced high; in fact, after taking up hunting, the student gave up drugs.

The essence of the hunting (generic here for hunting, trapping, and fishing) experience to me is the extreme focus it requires and engenders. Hunters must always be on the alert for their quarry, and that mental state automatically brings high anticipation and excitement.

My own hunting focus has been so extreme that, last November, it almost caused my undoing. I happen to devote my limited free time not to hunting but to trapping mink (I know, I know, that's terrible!). Having found a new mink stream, I was wading through one of those huge, ten-foot-square, concrete culverts that hold the road far above the stream. My hip boots sloshed with each step I took through the culvert in the foot-deep water. As I approached the open stream at the other end, my anticipation rose, for the banks of a wide pool in front of me were perfect for a mink set. With total focus directed at the bank ahead, I took one more step.

Suddenly, liquid ice engulfed me, instantly filling my boots and packbasket and stealing my breath. I was up to my armpits in the pool, having absentmindedly stepped off the end of the culvert. Truly I was at one with nature! Luckily, I was able to stay upright, wade to the bank, clamber up, shed what water I could, and return to the warmth of my vehicle.

Leopold would have understood the single-mindedness that led to my icy plunge. He recognized this type of focus in four categories of outdoorsmen: "the deer hunter habitually watches the next bend; the duck hunter watches the skyline; the bird hunter watches the dog; and the non-hunter does not watch" (Leopold 1953, 126). I am sure he would have been willing to add that "the mink trapper watches the stream bank."

Leopold's statement that "the non-hunter does not watch" precisely makes my point about the difference between the hunter and the nonexploitive outdoorsman. Hiking and camping and the like are wonderfully pleasant and enjoyable activities, and they certainly can inspire one toward conservation efforts. I cannot help but think, however, that if people have experienced the acute focus involved in the instinctive pursuit of natural quarry, as Aldo Leopold did regularly, their passion for wildlife conservation and working to inspire others in this endeavor will be greater.

"Instinctive" is the operative word here. With the human being's long evolutionary history of hunting, there should be little doubt that human genes encode some types of behavior that are only fully expressed during the search for prey. Furthermore, it is reasonable to conclude that, by behaving in ways our genes predispose us to behave, we will be rewarded by pleasure and passion.

In the modern world, however, almost all of the behavioral correlates of hunting, rather than being rewarded, are thwarted in most people. The rapid urbanization of society has left most folks devoid of opportunities to express and develop their natural hunting instincts. City toddlers, rather than bashing snakes, frogs, and all kinds of bugs, as they would have done in the countryside, are left with only ants to squash—if they are lucky. When older and at the nest-pilfering stage, most city kids are totally thwarted, and this continues throughout their development.

Is it any wonder, then, that by the time these urbanites have become thinking adults, most have forsaken their own nature? Never having been rewarded for practicing their natural instincts, they are either oblivious to the outdoors or afraid of it, or they view it with awe from the confines of their vehicles. The more daring ones, with perhaps some latent stirring of their instincts, may eventually take up hiking, backpacking, or other activities that allow them to interact more intimately with nature.

But I would argue that those who interact most intimately with the natural world are the hunters. I do not mean here the well-known "slob hunter" who steps afield a few days a year, shoots at anything, litters the woods, and gives real hunters a bad name. Rather, I mean the Aldo Leopold–type hunter, the one who takes a respectful, year-round interest in the natural world and for whom hunting is only one of many outdoor pursuits.

Hunting Skills and Rewards

Leopold became both a hunter and a naturalist at an early age, having been imbued with woods lore by his father, who himself was an accomplished hunter and conservationist (Meine 1988). Although Aldo Leopold did hunt with firearms, he also hunted with bow and arrows, which he made himself (McCabe 1987). This is a strong indication that Leopold felt, as any true hunter does, that it is the hunt itself, rather than the kill, that brings most of the satisfaction.

It is during the hunt when you need the acute focus and when anticipation and excitement pervade your being. But there is much more to the process than that. Reaching this point requires years of intimate interactions with the natural world, and the focus, excitement, and anticipation are the payoff as well as the motive for these interactions.

To begin with, you must gain as much knowledge about your quarry as possible and constantly strive for more. Next, by applying your natural alertness, you must develop the skill to detect your quarry and recognize its sign and habitat. Then, you must constantly test hypotheses about your quarry by casting a fly in the right pool, still-hunting on a chosen ridge, or setting a trap at a specific spot along a stream bank. You also need to hone your skill to interact appropriately with your quarry and to successfully apply your knowledge and experience in a wide variety of locations. Last, but not least, you must be able to withstand considerable adverse weather and field conditions in order to single-mindedly pursue your quest. "Who alone in our modern life so thrills to the sight of living beauty that he will endure hunger and thirst and cold to feed his eye upon it? The hunter" (Leopold 1953, 170).

Each of these components of the hunt greatly increases the intimacy with which you interact with the natural world, an integral part of hunting but not necessarily of the successful hiking, camping, or canoe trip. Because Aldo Leopold fished; hunted upland game, big game, predators, and waterfowl;

and trapped raccoons, coyotes, bobcats, and beavers (Leopold 1953), as well as other species, his intimate interactions with nature were considerable.

Leopold's personal rewards from his outdoor experiences, then, were also considerable. And that, I think, is what fueled his passion for conservation. Environmental passion, however, without a firm knowledge base to channel it, can lead to all kinds of unproductive or counterproductive results. As long as an issue is simple—"Save the Wolf"—passion alone can greatly benefit the cause. However, when things become complicated, like in determining how to manage the wolf once it has been saved, then uninformed passion can be detrimental.

I experienced this type of difficulty recently when I proposed that Minnesota's twenty-six hundred wolves would require population control. Soon after, I received an irate letter from a former supporter, who told me he had painstakingly scraped my autograph off a photo or print I had signed for him some years earlier.

Unlike the passion of those who are carried away by their feelings, Leopold's passion, of course, was much more highly informed. Leopold was trained in biology and forestry, but his long experience as a hunter and outdoorsman gave him valuable insights far beyond the book learning and theory of his formal training. Such a combination of knowledge and experience, although common in the environmental field a few decades ago, is becoming increasingly rare.

Today, many of the newer professional conservationists, especially those in academia, come from a variety of backgrounds including genetics, physiology, and economics. I know of a professor who teaches that truth is derived not just from science but also from intuition, ethics, and morals. To my mind, and by standard definition, he really means "perception," not "truth." Another conservation biology professor does not distinguish among recovery, reintroduction, and recolonization.

Even conservationists trained in fisheries and wildlife, if they have had no practical experience with fishing, hunting, or trapping and have never "felt the soil between their toes," are at a decided disadvantage in trying to understand the natural world, to teach about it, or to save it. This is only my opinion, of course, but I think that Aldo Leopold, the conservationist and hunter, would have agreed.

References

Flader, S. L. 1974. *Thinking like a mountain: Aldo Leopold and the evolution of an ethical attitude toward deer.* Columbia: University of Missouri Press.

Leopold, A. 1953. *Round River: from the journals of Aldo Leopold.* L. B. Leopold, ed. New York: Oxford University Press.

Leopold, A. 1966. *A sand county almanac.* New York: Ballantine Books.

McCabe, R. E. 1987. *Aldo Leopold the professor.* Amherst, Wisc.: Palmer Publications.

Meine, C. D. 1988. *Aldo Leopold: his life and work.* Madison: University of Wisconsin Press.

Chapter 12

The A–B Dichotomy and the Future

EDWIN P. PISTER

The A-B Dichotomy

With his inimitable combination of perception and prophecy, Aldo Leopold presented us in *A Sand County Almanac* with his "A-B cleavage," a phenomenon that has exerted great influence on the American conservation scene during the second half of the twentieth century. In "The Land Ethic," Leopold wrote, "Conservationists are notorious for their dissensions. . . . In each field one group (A) regards the land as soil, and its function as commodity-production; another group (B) regards the land as a biota, and its function as something broader" (Leopold 1949, 221).

We see this dichotomy across the spectrum of conservation professions. For example, a Group-A forester would be pushing to "get out the cut," paying minimal attention to environmental considerations. A Group-B forester would not necessarily oppose the proposed logging, but would be concerned that it be done only if riparian areas and endangered species were given proper protection. A Group-A fishery biologist's management goal might be to maximize sport catch in a new reservoir by introducing nonnative fishes, even when knowing they might spread throughout the drainage to the detriment of the native fishes. The Group-B fishery biologist would try to offer good fishing opportunities utilizing native species.

Probably nowhere has the A-B dichotomy been more pronounced and obvious than in the increasing concern exhibited over the protection of native biological diversity and related concepts concerning ecosystem management. Although the A-B phenomenon has long been noted in nonprofessional fish and wildlife circles (e.g., *Field and Stream* vs. *Audubon*), it also remains very pronounced within governmental resource management agencies, where those who present new concepts and innovative approaches often experience difficulties in gaining acceptance.

Background

In general, we find four distinct groups of professionals expressing concern over the conservation and integrity of native biodiversity (Pister 1992). First are those from the academic community (professors and students) who possess a deep appreciation of ecosystems and recognize the research potential within a biota (their professional existence may be strongly tied to its integrity). Unfortunately, such individuals (overwhelmingly Group B) tend to assume that conservation of species will be handled adequately under the stewardship mandate of conservation agencies and, therefore, seldom play an active role in biodiversity conservation.

In the second group are biologists from federal resource agencies such as the Fish and Wildlife Service, National Park Service, Bureau of Land Management, and Forest Service. Actions and attitudes relative to biodiversity conservation are determined primarily by the agency's basic orientation. For instance, a biologist representing the Office of Endangered Species of the Fish and Wildlife Service would probably be disturbed by a Corps of Engineers plan to dam a river critical to the existence of a threatened native plant or hydrobiid snail. Groups A and B are both represented in these agencies.

The third group is found within the private sector, such as at The Nature Conservancy, National Audubon Society, and private consulting firms. Nongovernmental conservation organizations are strongly supportive of biodiversity concepts (Group B), whereas private consulting firms often reflect the business circumstances under which they operate. Groups A and B will again be represented within the private sector.

In the fourth group, and providing perhaps the best example of Leopold's A-B dichotomy, are biologists within state fish and wildlife agencies entrusted with active management of most of the nation's fish and wildlife resources. Here, we often find a deep philosophical chasm separating biologists sitting at adjoining desks with supposedly similar educational and cultural backgrounds.

Why, then, does the A-B dichotomy exist throughout the fish and wildlife profession?

Why A-B?

I have observed that, historically, two basic types of biologists enter the fish and wildlife profession. First, there are those who develop an early love for fishing or hunting and pursue a related career by entering into a fish or wildlife curriculum at a college or university. Major specialty course requirements are supplemented with offerings designed primarily to sharpen technical skills. Foundation courses in the liberal arts, and theoretical courses such as environmental ethics, are avoided or minimized, and the student emerges at the bachelor's degree level perhaps *well trained,* but hardly *educated* in the classic sense (Baer 1978; Brown 1987). As employees, such individuals have a tendency to remain in Group A throughout their careers. They are technologically competent and, by reflecting agency policy (often with a strong Group-A bias), may rise quickly to administrative and policymaking levels.

Second, a type of student exists whose broad interest in nature causes him or her to major in the life sciences, often within a curriculum that requires strong grounding in the liberal arts. During the undergraduate years, the student develops an academic interest in fish or wildlife and finds that the best way to pursue this interest is through a career in a fish and wildlife agency. Such a person will seldom gain interest in fishing, hunting, or other consumptive uses. He or she may be viewed as something of an oddball by the old guard, whose allegiance is less to the entire biological resource than to a particular political constituency. Typically, this person will identify strongly with Group B and, when looking at his or her Group-A coworker at the next desk, will acknowledge a communication gap based on a very different set of values (Baer 1978).

Group-A employees will normally devote their careers to promoting traditional consumptive harvest programs of a fish and wildlife department, whereas the Group-B employee's primary concern will be with the fish and wildlife resource per se. There is a major difference between the two (Williams 1986).

A Personal Journey from Group A to Group B

Following the usual variety of moves, jobs, and agencies that accompany the earlier portion of most careers in fish and wildlife, I settled in the late 1950s into a position as a fishery biologist with the California Department of Fish and Game in the eastern Sierra Nevada Mountains and adjacent desert regions of the state (Pister 1985, 1987). My graduate studies in limnology had already given me knowledge of the geography of the area and insight into the biological characteristics of many of the waters falling under my jurisdiction. With one assistant, I was given the responsibility of managing nearly a thousand waters extending from the crest of the Sierra Nevada eastward to the Nevada state line and ranging from the top of Mt. Whitney, at 14,495 feet, to the floor of Death Valley, only eighty-one miles away from Mt. Whitney but nearly 282 feet below sea level. "Management," ideally, meant responsibility for the perpetuation of all species of aquatic organisms—including fishes, amphibians, invertebrates, and reptiles—and their habitats.

The management programs inherited from my predecessors reflected the philosophies of the times. They were, technologically, "state-of-the-art," and they were designed to meet the needs of a public hungry for outdoor recreation following World War II. They were, in short, model Group-A "utilitarian" management programs.

Aided by an efficient and well-funded fish hatchery system, most fishery biologists in California persevered in a heroic effort to increase catch-per-angler effort. With the exception of two trout (cutthroat and golden), the game fishes in my district were introduced species. Little was being done to ensure preservation of the native species because few people recognized what they were! Virtually no attention was devoted to the inventory and study of the basic components of the biota. We were living in a make-believe world. The California Department of Fish and Game, as was true of most state fish and wildlife agencies of that era, was spending its resources painting the building while the foundation showed signs of crumbling, as several species approached extinction. Although my department as a whole seemed pleased with what was going on, I felt a strange foreboding and knew that, somehow, things had to change. During the summer of 1964, I returned to *A Sand County Almanac* and reread "The Land Ethic" at leisure and in depth.

More than a decade of field experience had given Leopold's words new meaning. Within the principles he eloquently set forth, I found a rational basis for approaching and solving the problems that had perplexed and overwhelmed me. I felt I had within my grasp the basic components for making management programs address the entire biota, not simply the popular demands for fishing that had up until then consumed my time.

A Sand County Almanac, and the concept of the A-B dichotomy, confirmed my observation that, to be really meaningful and to serve the long-term interests of the biota (and, therefore, the people), management programs should begin with the integrity of the land and water. Using this as a foundation, the resource manager should then build a pyramid upward, first adding a basic concern for the flora and fauna and finally adding species of special economic and political interest. I had toyed with this possibility in the past but had been unable to muster the courage necessary to challenge the system. Leopold's grit, as well as his clear purpose and simple means, seemed to be the very thing I needed to gain this courage. I was especially motivated by his wry observation that "nonconformity is the highest evolutionary attainment of social animals" (Leopold 1953, 8). In retrospect, the motivating force in my change from Group A to Group B was Aldo Leopold's summary precept: "A thing is right when it tends to preserve the integrity, stability, and beauty of the biotic community. It is wrong when it tends otherwise" (Leopold 1949, 224–225). It became clear to me that one could not support this precept and remain in Group A.

The universe is governed by the complex interaction of immutable and elegant physical laws. As I read through *A Sand County Almanac,* I realized that a set of natural laws, equally elegant and immutable, governed biological systems as well. The futility of trying to circumvent these laws for any appreciable length of time in the service of short-term economic and political goals became even more apparent.

In the early 1960s, the nation was experiencing the initial throes of a reawakened conservation movement, precipitated by publication of Rachel Carson's *Silent Spring* (1962). The first Endangered Species Act was not passed until 1966, and the National Environmental Policy Act of 1969 was still several years away. Financial support for innovative programs hardly existed, and administrative (and public) backing was similarly lacking. The land ethic was not yet a force to be reckoned with. Nevertheless, it seemed unconscionable to me for my department to be spending the majority of its fisheries budget planting put-and-take trout while ignoring the indigenous biotic community. When the pressing need to conduct biological surveys was brought to the attention of our top administrators, even those with advanced academic training responded with blank looks or remarks to the effect that the public and the legislature would never allow license money to be spent collecting "bugs or suckers," some of which we were vigorously attempting to eradicate to provide better angling.

Adding insult to injury, the department commissioned several fine wildlife scholars and administrators (headed by Starker Leopold, Aldo's eldest son and my major professor at the University of California–Berkeley) to prepare a plan to guide the department's activities until 1980. The principles of the plan were ecologically sound and, had they been followed, would have done much to reverse the status quo. However, politics and tradition spoke louder than logic, and we stumbled on as if the plan had never been written.

The sense of futility in continuing to blindly repeat conventional management procedures caused me to look elsewhere for long-term solutions to problems that were growing worse with each passing day. The dichotomy between Group A and Group B is rarely absolute. Most of us embody elements of both groups in a balance that tends to change through the years, often as a result of new perceptions and maturing values. I believe that most scientists, in the course of their careers, tend to move from Group A toward Group B and seldom vice versa. Leopold's land ethic served to establish the ideal that I and others strived to achieve: habitat integrity and a complete complement of native species. *A Sand County Almanac* literally charted our course: "A land ethic of course cannot prevent the alteration, management, and use of these 'resources,' but it does affirm their right to continued existence, and, at least in spots, their continued existence in a natural state" (Leopold 1949, 204).

One could sense the beginnings of change, although our management and research programs remained utilitarian. Actually, "economically or politically expedient" would better characterize them, because enlightened self-interest (the hallmark of utilitarianism), if carried to its logical limits, must lead to a policy of basic resource integrity and protection.

With the blessings of an understanding supervisor, I began to substitute species inventories for creel censuses. The initial survey revealed that three of the four fishes native to the Owens River system were either endangered or of indeterminate status. Eventually, informal species recovery plans and similar nonconsumptive management programs were initiated, along with a plan to promote populations of self-sustaining wild trout. Major efforts were devoted to environmental protection. Additional efforts were undertaken to educate key department personnel and the general public and to win the political support necessary for the development of an environmentally sound program of resource management. It is gratifying to note that this new program was generally successful and did not, as initially feared, provoke significant adverse public or political reaction.

Leopold's A-B cleavage persists in the fish and wildlife profession. Group-B types, who view land as an integrated biota, still identify strongly with academia, although their ranks within the agencies are growing rapidly. Group-A types, who regard the land primarily as a vehicle for producing a harvestable crop, remain prominent within the agencies, especially at the upper echelons. The love of fishing and hunting that initially brought them into the profession remains their primary motivation, and their management goals frequently have not progressed much beyond that. The significance of this is often overlooked by analysts seeking to understand the history and evolution of the nation's fish and wildlife programs. These Group-A types were quick to discover that license buyers, who fund most agency programs, were also overwhelmingly of the Group-A type, so it was easy to obtain the legislative, administrative, and financial backing consistent with economically popular programs. Thus, the ecologically essential nongame component of the biota was usually neglected (and often disdained) in favor of economically valuable species. To paraphrase Leopold, we, in the agencies, fancy that game species support us, forgetting what supports game species (Leopold 1949, 178). This support is manifested in healthy land and a healthy, intact biota: soil, plants, invertebrates, nongame and game species alike. My own professional guidepost remains Leopold's summary precept of right and wrong within the biotic community (Leopold 1949, 224–225).

It is a credit to Aldo Leopold that increasing concern exists today for the wildlife that share common ecosystems with us. More and more, scientists and our diverse publics think in terms of species and ecosystems in efforts to preserve biological diversity.

The Future

Although Group-A types still exert a strong influence within fish and wildlife agencies, their numbers will inevitably decrease through attrition as the old guard is replaced by younger workers educated in a more enlightened era. Even institutions formerly labeled as "trade schools" (which produced much of the nation's post–World War II fish and wildlife leadership) are now updating their curricula to embrace the biological sciences in a broader sense. Group-A-oriented courses are still offered, but normally within the context of a Group-B educational process.

German physicist Max Planck (1950) wryly observed: "A new scientific truth does not triumph by convincing its opponents and making them see the light, but rather because its opponents eventually die, and a new generation grows up that is familiar with it." To be fair to those Group-A types that remain, this process must occur every generation or two because humans function this way. The future of the nation's fish and wildlife resources will depend on our ability to recognize and admit our past mistakes and failures and to structure programs so as to avoid repeating them.

I feel we are heading in the right direction. Group-A types reflected the utilitarianism that predominated before and after World War II and were trained in schools whose fish and wildlife curricula were structured accordingly. They usually approached their work with the best of intentions and fulfilled their job assignments well, but often with little concern for ecological consequences. It was only during the environmental era beginning in the 1960s that Group-B types began to emerge in force.

In the final analysis, the educational process will expedite the necessary transition from Group A to Group B. Aldo Leopold (1953) observed in *Round River* that one of the penalties of an ecological education is that one lives alone in a world of wounds. As tomorrow's fish and wildlife students emerge from our colleges and universities, more aware of ecological principles and supported by an increasingly well-informed electorate, they may come to share with us the wounds inherent in programs built around stocked nonnative trout and introduced game species. And as these wounds are felt and shared throughout the profession, we may begin to view Group A as a remnant of a past era while we proceed into this new century guided by a new ethic built around what we can do for our fish and wildlife resources, rather than what they can do for us.

References

Baer, R. A. 1978. Values in higher education: the crisis and the promise. *Journal of the National Association of Colleges and Teachers of Agriculture* March:4–10.

Brown, D. A. 1987. Ethics, science, and environmental regulation. *Environmental Ethics* 9:331–349.

Carson, R. 1962. *Silent spring*. Boston, Mass.: Houghton Mifflin Company.

Leopold, A. 1949. *A sand county almanac and sketches here and there*. New York: Oxford University Press.

Leopold, A. 1953. *Round River: from the journals of Aldo Leopold*. L. B. Leopold, ed. New York: Oxford University Press.

Pister, E. P. 1985. Desert pupfishes: reflections on reality, desirability, and conscience. *Environmental Biology of Fishes* 12:3–12.

Pister, E. P. 1987. A pilgrim's progress from group A to group B. In *Companion to* A Sand County Almanac: *interpretive and critical essays*, ed. J. B. Callicott, 221–232. Madison: University of Wisconsin Press.

Pister, E. P. 1992. Ethical considerations in conservation of biodiversity. In *Biological diversity in aquatic management*, ed. J. E. Williams and R. J. Neves, 355–364. Washington, D.C.: Wildlife Management Institute.

Planck, M. 1950. Scientific autobiography and other papers. In *Respectfully quoted: A dictionary of quotations from the Library of Congress*, ed. S. Platt (1992), 33–34. Washington, D.C.: Congressional Quarterly, Inc.

Williams, T. 1986. Who's managing the wildlife managers? *Orion Nature Quarterly* 5(4):16–23.

Chapter 13

What Would Aldo Have Done?

A Personal Story

JACK WARD THOMAS

This essay takes on, of necessity, a very personal tone. I do not fancy myself a Leopold scholar. I have developed a sense of knowing the man only through his writings; through the writings of his children, friends, and students; and through those same people whom I have had the honor to know. These feelings of knowing Aldo Leopold and contacts are personal to me.

First Encounter

My first encounter with the works of Aldo Leopold took place in 1956, during my senior year at the Agricultural and Mechanical College of Texas. I was enrolled in a course called, as I remember, "Big Game Management" (which says something about the status of the wildlife management profession at the time). Our teacher was Dr. O. Charles ("Charlie") Wallmo, who had been a student at the University of Wisconsin in the department established by Aldo Leopold. The primary text in that course was Leopold's *Game Management* (1933).

The name Aldo Leopold meant nothing to a young man from rural Texas who thought, in the absence of any better idea, that it would be a marvelous thing to make a living out of one of two primary interests so far evidenced in life—hunting and fishing. I was, however, fascinated by that book. I read it several times. That was more attention than I had ever paid to any book, much less a textbook. By today's standards, there was little "science" behind the wildlife profession in those days. There was even less science when that book was first published in 1933.

But the author had taken what science there was and patched together a text for a fledgling profession with the glue of keen observation, common sense, and intuition. That was one heck of a book in 1933. It was still the best text on the subject in 1956. And it remains a remarkable book even today.

Much of the research in the game management business over the decades that followed its publication centered on testing the hypotheses and the principles of wildlife management that Leopold put forth. It was, without doubt, a seminal work and a remarkable stimulus to the budding wildlife management profession. For example, in 1979, several colleagues and I put forth an elk habitat optimization model predicated primarily on elk habitat use relative to the edges between openings and forest cover. This led to a description of forest stand sizes and openings and their arrangement in time and space. This model, which received wide use for a time, came to be called the "60/40 Optimization Model" for elk habitat (60% in openings and 40% in cover).

Sometime later, I happened to be browsing through *Game Management* one stormy winter night while lounging in front of the wood stove, and I found that Leopold had come to a similar conclusion based on observation and intuition—some forty-two years earlier! Had I (we) forgotten that? Or, more likely, were we having difficulty knowing where our thinking, observations, and intuitions began and those of Aldo Leopold ended? This was not the first or the last such occasion in my career.

A Sand County Almanac—A First Contact

Late in that course on big game management, Charlie Wallmo assigned our class the chore—and we did think of it as a chore—of preparing individual book reviews of a small volume titled *A Sand County Almanac* (Leopold 1949), authored by that same Aldo Leopold. This was in 1956, only seven years after the book's publication. At that time, *A Sand County Almanac* had not yet been designated a "classic," as it was so overwhelmingly referred to in later years.

There were eight of us, all neophytes, steeped in the concept that wildlife management was a purely scientific endeavor unsullied by intrusions of feeling, sentiment, or philosophical consideration, panned that book. Each of us, more or less independently, saw nothing in that book of any interest or use to us budding, cold-blooded, rational technocrats. And then, adding insult to injury, we continued to pound home those opinions during the ensuing class discussions.

When I think back on that day, I wonder at Dr. Wallmo's foresight in introducing us to *A Sand County Almanac* and *Game Management* and, through those books, to Aldo Leopold. As I recall, none of us made any connection between the pragmatism of *Game Management* and the philosophical, moral, and ethical aspects of *A Sand County Almanac*—the two sides of the coin of conservation. I marvel at the restraint that Charlie Wallmo showed to the swine trampling on the pearls that he had placed before us. He was too shrewd for anger or disappointment to show. He knew that we were not yet ready to absorb the wisdom and insights of *A Sand County Almanac.* But I think now that he knew—or hoped—that someday we would be ready.

And so we were planted with seeds that would lie dormant until circumstances were right to support growth in mind and soul. This was the kind of growth that would continue for a lifetime, with the race for sunlight and with the growth rings becoming wider with each year of deepening and spreading roots. Perhaps this is what also happened to Leopold in the time between when he wrote *Game Management* and when he penned the essays that emerged in *A Sand County Almanac.*

Several years after achieving gainful employment in the business of game management with the Texas Game and Fish Commission, I had occasion to retrieve *A Sand County Almanac* from the box where it had been packed away. I searched the book looking for the context of a quotation that I had encountered in an article that piqued my curiosity. That induced me to read *A Sand County Almanac* again—and again. The essays made more sense and "tracks" began to appear in the margins and white spaces of my copy—underlines, comments, questions, and observations. The freshest tracks in that same volume appeared only last week.

The Leopold Connections—Many and Varied

At that time in my life other influences provided more connections to Aldo Leopold. My first boss in the old Texas Game and Fish Commission was

James G. Teer. He had already finished half of his course work for his doctorate at the University of Wisconsin in the department established by Leopold. His primary advisor was Robert McCabe, one of Leopold's first graduate students, later a colleague, and then Leopold's successor as department head at the University of Wisconsin.

Because I had no real idea of what a wildlife biologist was supposed to be or do, I imprinted on Jim Teer in much the same way that a newly hatched gosling imprints on the first moving thing it sees. In this case, the imprinting was most appropriate, for he was the epitome of a wildlife biologist. Even today, I cannot think of a better mentor or colleague.

Not surprisingly, Jim Teer was big on discussing the writings and philosophy of Aldo Leopold as we sat around campfires or in the cabins that we occupied during field seasons (motels were considered a bit "spendy" in those days). My immersion into the Leopold mystique was reinforced by several visits to our research operations (which were intended to produce Teer's doctoral dissertation) by Professors Robert McCabe and Joe Hickey of the University of Wisconsin.

The young biologists and technicians and I were fascinated by the conversations that took place between McCabe, Hickey, and Teer as they sat in the game department cabin known as Guthrie Camp, near the headwaters of the Llano River. They discussed Leopold the man, the professor, the philosopher and sage, and his actions and style. Interwoven into these conversations were the resource management questions of the day. In the midst of such a conversation, Dr. McCabe interrupted with a question: "Let us consider. What would Aldo have done?"

To the young philistines who were seated outside the inner circle and listening respectfully to the discourse, that question struck us as a bit pompous, and certainly a tad theatrical. We watched to see if this was a gentle joke. It was not. McCabe was deadly serious. We found all of this quite amusing and we exchanged sideways glances—though we dared not smile. For months afterward, when some serious question arose, one of us would stroke our chin, cast our eyes to the heavens, and ask in sepulchral tones—"What would Aldo have done?"

As the years passed, the joking refrain of "What would Aldo have done?" took on a more and more serious tone. And it became, at least for me, a point of reference from which to face ethical or philosophical challenges.

Impressions for a Lifetime

Some two or three years into my professional career, I had my first chance to attend a North American Wildlife and Natural Resources Conference. I was much impressed when, at the closing banquet, the president of The Wildlife Society presented the Aldo Leopold Medal, the ultimate recognition of a wildlife biologist by his peer group. I could not conceive, then or now, of any greater honor than being elected by colleagues to the presidency of The Wildlife Society or being selected as the recipient of the Leopold Medal. Knowing that Leopold was one of the founders and an early president of the society added to the feeling of the moment.

That experience led me to seek out other Leopold writings, such as *Round River: From the Journals of Aldo Leopold* (Leopold 1953) and The River of the Mother of God *and Other Essays by Aldo Leopold* (Flader and Callicott 1991), and to read, again and again, *A Sand County Almanac.* Ever since then, when I pick up *A Sand County Almanac,* I am reminded of a passage in the poem "Little Gidding" from T. S. Eliot's *Four Quartets*:

> We shall not cease from exploration
> And the end of all our exploring
> Will be to arrive where we started
> And know the place for the first time.

In a sense, every time I return to Leopold's writings, it is to arrive where I started and to see the world for the first time. In the course of that latest visit, I discern things in the written landscape that I had not sensed before. This, I believe, is the result of things that I have learned and seen and felt during my explorations since my most recent arrival at the beginning. Colleagues have expressed similar feelings to me. These feelings seem to have been enhanced over the years as personal experience, sensitized perception, and thoughts melded synergistically with the words of Aldo Leopold.

Some forty-two years have passed since those days when I listened to McCabe, Hickey, and Teer discuss Aldo Leopold and his writings late into the gentle nights of the Texas Hill Country. As chance would have it, over ensuing decades, Bob McCabe, Jim Teer, and I were each honored by election to the presidency of The Wildlife Society. And, along with Joe Hickey, we were each presented with the Leopold Medal.

Even today, when I see my old friend and mentor, Jim Teer, we are apt to engage in discussion of some weighty matter or another. And one of us is apt to ask in that joking way adopted so long ago, "What would Aldo have done?" As is common in older male Texans, deep feelings are often masked by attempts at humor.

Learning More from Historians

My interest in Leopold the man was much enhanced by reading Curt Meine's (1988) *Aldo Leopold: His Life and Work,* Robert McCabe's (1987) *Aldo Leopold: the Professor,* Marybeth Lorbiecki's (1996) *Aldo Leopold: A Fierce Green Fire,* and Susan Flader's (1977) *Thinking Like a Mountain: Aldo Leopold and the Evolution of an Ecological Attitude toward Deer, Wolves, and Forests.* In these books, I began to recognize Aldo Leopold as a man who had been formed by experiences similar to mine, and perhaps to those of most natural resource professionals. The challenges faced by the wildlife profession seem to change only slightly with time and place—and to reappear again and again, almost as specters.

Leopold was not a superman. He made mistakes—some quite significant mistakes (Meine 1988). Some of his early writings were varyingly maudlin, simplistic, strident, and marked with the hubris of youth. He was a greenhorn once and made his share of greenhorn mistakes. I think he would have encouraged discovery through such mistakes. He would, I think, have been distressed by deification.

He had a wife and a remarkable family for whom he cared deeply. He was sometimes confused about his path in life and surely a bit frightened from time to time. He was involved in professional disagreements and some outright fights. He was not universally admired—not during his lifetime and not even today (Meine 1988).

But he never quit. He never quit thinking and rethinking. He kept writing—in both popular and technical styles. And, unlike most people, he dared to seek a greater approximation of truth and larger meanings garnered from everyday experience. He was not afraid to stretch out from a technical background to deal with philosophy and ethics. He came to see that humble beginnings in game management, forestry, and agricultural husbandry could lead to a greater appreciation, and even reverence, for the good earth and its web of life.

When Neil Armstrong stepped out of the lunar lander onto the surface of the moon, he said, "That is one small step for man and one giant leap for mankind." The most lasting value of that moment was not the technological achievement of a man stepping onto the surface of the moon. The most lasting value, for me at least, was the image beamed back to Earth of our beautiful blue and white and brown world spinning against the cold blackness of space. How beautiful, how incredibly rare, and how vulnerable seemed our home. That picture forever changed my view of the Earth and of my place on that Earth—as it did for millions.

Yet, I sense today that Aldo Leopold had experienced this same philosophical epiphany a full fifty years before Neil Armstrong took that "giant leap for mankind." Leopold's giant leap is documented in the essay "The Land Ethic" in *A Sand County Almanac.*

Honors Bestowed—Better Late than Never

Aldo Leopold spent his early professional career with the then fledgling U.S. Forest Service—as did a number of other outstanding pioneers in natural resource management. When I arrived in Washington, D.C. to take on Gifford Pinchot's mantle as the thirteenth chief of the Forest Service, I went into the headquarters building on a Sunday afternoon before my duties began on the following day. As I wandered through the old brick building, I found the Chief's Conference Room (which was once the office of Gifford Pinchot, founder of the Forest Service). The hallway that led to that room contained the photographs of the twelve chiefs that had preceded me since 1905. I looked into the eyes of each of the chiefs and strongly felt the traditions, the responsibilities, and the opportunities. Yet, something was missing from that hallway of honor and tradition. I felt strangely dissatisfied.

On Monday morning, I issued my first instruction as chief. That order was to place in that hallway the photographs of Arthur Carhart, Robert Marshall, and Aldo Leopold. These were the individuals of intellect, vision, and soul who had as much to do with the evolution of the Forest Service as any chief—including Gifford Pinchot. To do this was my obligation and my honor—and, to my mind, it was long overdue.

My second act was to begin work on *The Forest Service Ethics and A Course to the Future* (Thomas 1995). That document laid out, for the first time, a land-use ethic for Forest Service personnel to rely on in the management of some 194 million acres of national forests and grasslands. These are the lands of all the American people and a legacy unique in all the world. It was my intent that we take up Leopold's call of forty-five years earlier for the use of a land ethic in management. That statement read: "Our land ethic is to: Promote the sustainability of ecosystems by ensuring their health, diversity, and productivity" (Thomas 1995, 1). Such was my obligation and my honor—and it was long overdue.

The Leopold Medal—Honor and Reminder

During the years I served as chief of the Forest Service, the bronze Aldo Leopold Medal sat on my desk in the Chief's Office in the nation's capital. Aldo Leopold's image is on that medal. The medal did not face out toward the room so that others could see it. It faced my desk chair so that I could see the medal and the image of Aldo Leopold. It was not so positioned to remind me of the high honor bestowed upon me by colleagues and cherished friends. The medal faced me as a constant reminder that one can gain wisdom by searching and growing intellectually and spiritually with age and experience. And it reminded me that there are such things as honor and hope and duty. Most important of all, the presence of that medal enticed me to ask, when there were tough decisions to be made or attacks to be endured, "What would Aldo have done?"

Thinking Like a Mountain—
Things Do Change

Today, in my new career as a professor at the University of Montana, I keep the medal in my desk drawer to remind me to temper my teaching of technical matters with what passes for wisdom and some philosophical intonations from Leopold and others. And, as Charlie Wallmo did for me in class some forty-three years ago, I place a copy of *A Sand County Almanac* in the students' hands and ask for their reviews. These reviews differ from those made in my class so long ago at Texas A&M. There is every indication that students today can think and feel beyond the technical. It is, I think, a sign of progress and maturity in our profession—students beginning to "think like a mountain." That, indeed, warms my heart. I am certain that Dr. Wallmo would have been pleased to see how the seeds he sowed have grown.

References

Flader, S. L. 1977. *Thinking like a mountain: Aldo Leopold and the evolution of an ecological attitude toward deer, wolves, and forests.* Columbia: University of Missouri Press.

Flader, S. L., and J. B. Callicott, eds. 1991. The River of the Mother of God *and other essays by Aldo Leopold.* Madison: University of Wisconsin Press.

Leopold, A. 1933. *Game management.* New York: Charles Scribner's Sons.

Leopold, A. 1949. *A sand county almanac and sketches here and there.* New York: Oxford University Press.

Leopold, A. 1953. *Round River: from the journals of Aldo Leopold.* L. B. Leopold, ed. New York: Oxford University Press.

Lorbiecki, M. 1996. *Aldo Leopold: a fierce green fire.* Helena, Mont.: Falcon Press.

McCabe, R. A. 1987. *Aldo Leopold: the professor.* Amherst, Wisc.: Palmer Publications.

Meine, C. 1988. *Aldo Leopold: his life and work.* Madison: University of Wisconsin Press.

Thomas, J. W. 1995. *The Forest Service ethics and a course to the future.* Washington, D.C.: USDA Forest Service.

Index

Page numbers followed by *t* refer to tables.